VIRTUAL WORK AND ENE[RGY]

COLIN J. ROWE, B.A.
193, WATLING STREET, GRENDON,
Nr. ATHERSTONE, WARWICKS,
ENGLAND, CV9 2PJ
Phone: ATHERSTONE 4891

Other titles in the series

✱

Force Systems and Equilibrium
J. G. A. CROLL

Stress and Strain
T. R. GRAVES SMITH

Simple Bending
J. R. TYNE

Torsion
A. C. WALKER

The Buckling of Struts
A. C. WALKER

Modular Textbooks in Engineering

Editor: A. C. WALKER, Ph.D.
*Department of Civil and Municipal Engineering,
University College London*

Virtual Work and Energy Concepts

J. RHODES
B.Sc., Ph.D.
*Department of Mechanics of Materials
University of Strathclyde*

1975

CHATTO & WINDUS

LONDON

Published by
Chatto & Windus Ltd
42 William IV Street
London WC2N 4DF

*

Clarke, Irwin & Co. Ltd
Toronto

All rights reserved. No part of this publication may be reproduced, stored in a retrieval system, or transmitted, in any form, or by any means, electronic, mechanical, photocopying, recording or otherwise, without the prior permission of Chatto & Windus Ltd

ISBN 0 7011 1951 9

© J. Rhodes 1975

Printed in Great Britain by
T. & A. Constable Ltd
Hopetoun Street
Edinburgh EH7 4NF

CONTENTS

1 Virtual work *page* 7
1.1 Introduction 7
1.2 Work done by a force 8
1.3 The principle of virtual work (virtual displacements) 8
1.4 The use of virtual work for the analysis of rigid member structures 10
Problems 13

2 Strain energy 15
2.1 Introduction 15
2.2 Virtual work for an elastic body 16
2.3 Systems with many degrees of freedom 22
2.4 The principle of virtual complementary work 26
2.5 The unit load method 27
2.6 Castigliano's theorem 29
Problems 31

3 Deflections of statically determinate structures 34
3.1 Introduction 34
3.2 Formulation of virtual complementary energy expressions for various loading conditions 35
3.3 Deflections of pin-jointed frames 38
Problems 41

4 Deflections of beams 43
4.1 Introduction 43
4.2 Deflections of beams 43
4.3 Structures with several beam and bar members 47
Problems 50

CONTENTS

5 Analysis of statically indeterminate structures   53
   5.1 Statically indeterminate beams   53
   5.2 Statically indeterminate frameworks   59
   5.3 Deflections of statically indeterminate structures   64
     Problems   66

Answers to problems   69

Index   71

1

VIRTUAL WORK

1.1 Introduction

In engineering it is essential to know how bodies such as machine components or structures behave under load; how they deform, what internal forces and actions are set up, how the supports react to applied load. For many years engineers have faced these problems and through their efforts and ingenuity the science of structural mechanics has been developed.

In the development of the capacity to analyse structural problems, two main methods of approach have been formulated. The first follows directly from the fundamental rule of mechanics, *the principle of equilibrium*. Using this approach, direct application of the equations of static equilibrium are employed to analyse the problem. For many simple problems the equations of equilibrium by themselves provide a unique solution. For more difficult problems the analysis is dependent on the deformations and displacements of the body under consideration. In these cases the principles of equilibrium are applied to small elements of the body, and used in conjunction with the load-deformation relationships of the material to formulate differential equations from which the problem can be solved. (An example of this is the beam deflection equation $EI \frac{d^2y}{dx^2} - M_x = 0$, see, for example, J. R. Tyne, *Simple Bending*, in this series.)

The second method of approach, which is the subject of this book, uses the concepts of work and energy to analyse these problems and follows from *the principle of virtual work*. Although this method does not have quite so long a history as the equilibrium approach, it is by no means a recently discovered idea. Leonardo da Vinci (1452-1519) was aware of the idea of applying virtual displacements to bodies and applied it to the analysis of pulleys and levers. Galileo (1564-1642) and Stevin (1548-1620) further developed the concept of virtual work, but John Bernoulli (1647-1748) is credited with the first statement of the most general form of the principle of virtual work. From these beginnings the findings of later great researchers such as Lagrange (1736-1813), Clapeyron (1799-1864), Castigliano (1847-1884), Mohr (1835-1918), Engesser (1848-1931), and many others have developed the use of energy principles to such an extent that in many cases the use of this approach is by far preferable to the approach based directly on equilibrium principles.

In this chapter the principle of virtual work is derived in its simplest form and is applied to the analysis of rigid member structures. In succeeding chapters, the more general form of this principle is developed, together with some of the energy theorems which are derived from the basic principle.

1.2 Work done by a force

The work done when a force moves its point of application is defined as the product of the force magnitude and the distance moved by its point of application in the direction of its line of action. This is illustrated in Fig. 1.1. Here the point of application of the force F moves from A to B,

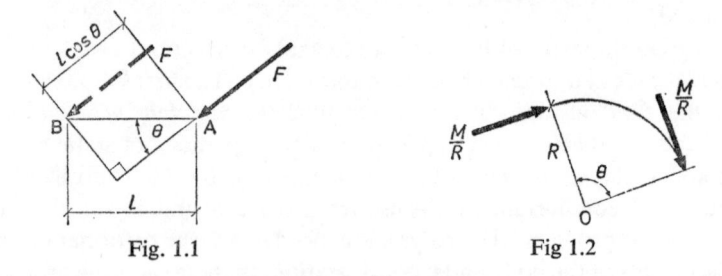

Fig. 1.1

Fig 1.2

a distance l at an angle θ to the line of action of F. The work done by F is its magnitude times the distance moved along its line of action,

$$W \equiv Fd = Fl \cos \theta. \tag{1.1}$$

The work done when a moment rotates its point of application about a fixed point in space is found by considering the moment as a force applied at a radius R from the point as shown in Fig. 1.2. By definition the force required to produce a moment M is equal to M/R. The distance moved by the force due to a rotation about the point O is equal to $R\theta$. The work done by the moment is therefore

$$W \equiv \frac{M}{R} \times R\theta = M\theta. \tag{1.2}$$

1.3 The principle of virtual work (virtual displacements)

Virtual work, as distinct from real work, is performed when a real force moves through a *virtual displacement*. A virtual displacement of a body is any possible arbitrary displacement we can imagine the body to undergo within the limitations imposed by any external constraints. Thus a virtual displacement is not a displacement which the body undergoes as a result of the actual loading, but is an imaginary displacement which the body could possibly undergo as a result of some entirely arbitrary, and in general unknown, loading.

In its simplest form the principle of virtual work follows from a consideration of the virtual work done by forces acting on a particle when the particle is given any arbitrary virtual displacement as shown in Fig. 1.3.

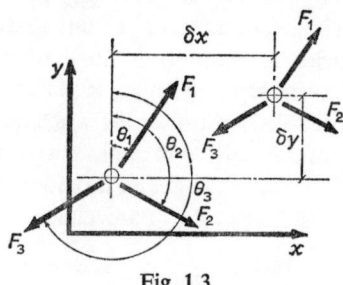

Fig. 1.3

The net virtual work done during the virtual displacement is found by summing the virtual work done by each force,

$$\delta W = F_1 \sin \theta_1 \delta x + F_2 \sin \theta_2 \delta x + F_3 \sin \theta_3 \delta x$$
$$+ F_1 \cos \theta_1 \delta y + F_2 \cos \theta_2 \delta y + F_3 \cos \theta_3 \delta y$$
$$= R_x \, \delta x + R_y \, \delta y, \qquad (1.3)$$

where $R_x = \Sigma F \sin \theta$ is the resultant force in the x direction and $R_y = \Sigma F \cos \theta$ is the resultant force in the y direction.

Now let us examine the conditions under which the virtual work is zero. Obviously for given values of R_x and R_y we can obtain particular virtual displacements which will result in zero virtual work. However, the only condition under which zero virtual work will occur under an *arbitrary* virtual displacement (i.e. arbitrary values of δ_x and δ_y) is that R_x and R_y are both zero. Now if R_x and R_y are zero, the particle is in equilibrium under the forces acting on it and the virtual work expression becomes

$$\delta W = 0. \qquad (1.4)$$

We can generalise this example by stating that 'If the net work done by a system of forces acting on a particle is zero during an arbitrary virtual displacement, then the particle is in equilibrium'. This is the formal statement of the principle of virtual work, also called *the principle of virtual displacements,* since it is obtained by the application of virtual displacements.

Note that the magnitude of the applied virtual displacement is arbitrary. It is very useful in some problems to make the virtual displacements infinitesimally small, as we shall see in the problems encountered in this chapter.

1.4 The use of virtual work for the analysis of rigid member structures

In this section we consider the application of the principle of virtual work to the equilibrium analysis of structures whose members may be considered rigid, i.e. they do not deform in any way under load.

Each member of such a structure can be considered to be made up of a large number of particles in equilibrium so that the virtual work done during a virtual displacement of the member is zero simultaneously for each particle and we need only consider the work done by external loads.

As a simple example of the method of approach, consider the equilibrium of the lever shown in Fig. 1.4 on which an unknown force P is required to balance specified loads W_1, W_2, W_3. To evaluate P we give the lever an

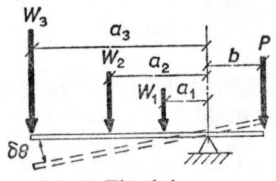

Fig. 1.4

infinitesimally small virtual rotation $\delta\theta$ about its fulcrum. The distance moved by each load is given by the product of $\delta\theta$ and the distance of the load from the fulcrum. Therefore

$$\delta W = (W_1 a_1 + W_2 a_2 + W_3 a_3 - Pb)\delta\theta.$$

Note that the work done by P is negative since the virtual displacement of its point of action is opposite in direction to P.

Equating δW to zero furnishes the value of P.

$$Pb = W_1 a_1 + W_2 a_2 + W_3 a_3.$$

This equation is identical to that which could be obtained by taking moments about the fulcrum. There is therefore little benefit to be gained from the use of virtual work for simple problems such as this, since the solutions can be obtained just as easily by direct application of the equations of equilibrium.

If we consider systems of members linked together to form mechanisms, however, then in many cases the method of virtual work can perform the required analysis very much more quickly than the direct application of equilibrium principles. This is mainly because we do not need to concern ourselves with reactions which do not move and thus perform no work during a virtual displacement.

VIRTUAL WORK

In this book we shall consider only *ideal systems*. These may be defined as systems in which no work is dissipated by friction and for which equation (1.4) holds for the virtual work done during a virtual displacement. The application of virtual work to this type of structure is illustrated by the following examples.

Example 1.1

The mechanism shown in Fig. 1.5 consists of three uniform members of length a and weight W pinned at their junctions and at the supports A and B. Obtain the value of P required to hold the mechanism in position in terms of W and θ.

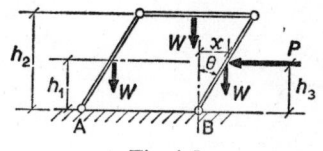

Fig. 1.5

Solution. The weights of the members may be assumed to act through their centroids, i.e. at their mid-points.

The heights of the centroids of the members are

$$h_1 = h_3 = \tfrac{1}{2}a \cos \theta, \quad h_2 = a \cos \theta.$$

Now if a virtual rotation $\delta\theta$ is applied to the mechanism, the change in height of the centroids can be obtained by differentiating the expressions for h_1 and h_2 to obtain δh_1 and δh_2,

$$\delta h_1 = \delta h_3 = -\tfrac{1}{2}a \sin \theta \, \delta\theta, \quad \delta h_2 = -a \sin \theta \, \delta\theta.$$

The horizontal distance δx moved by the load is found by putting x in terms of θ and differentiating,

$$x = \tfrac{1}{2}a \sin \theta, \quad \delta x = \tfrac{1}{2}a \cos \theta \, \delta\theta.$$

The virtual work done during the virtual displacement $\delta\theta$ is

$$\delta W = -(W \delta h_1 + W \delta h_2 + W \delta h_3 + P \, \delta x).$$

(Note that the negative sign indicates that the loads and force P do negative work if δh and δx are positive, i.e. if h and x increase, then the displacements are in the opposite direction to the applied loads.) Substituting for δh and δx gives

$$\delta W = W(\tfrac{1}{2}a \sin \theta \times 2 + a \sin \theta) \, \delta\theta - P\tfrac{1}{2}a \cos \theta \, \delta\theta.$$

Now, since for equilibrium $\delta W = 0$ we have

$$2Wa \sin \theta - P\tfrac{1}{2}a \cos \theta = 0.$$

Therefore $P = 4W \tan \theta$.

This problem shows the advantage which can be gained by using virtual work for equilibrium problems. If direct application of the equilibrium

VIRTUAL WORK AND ENERGY CONCEPTS

equations had been used, the structure would have been dissected and each part investigated individually.

Example 1.2

Two slender weightless rods AB and CDE have forces P and W at C and B as shown in Fig. 1.6. AB is hinged at A and CDE at D. The roller at E is negligibly small and allows frictionless relative movement of the two rods at this point. Find the angle θ for equilibrium in terms of W, P, l and a.

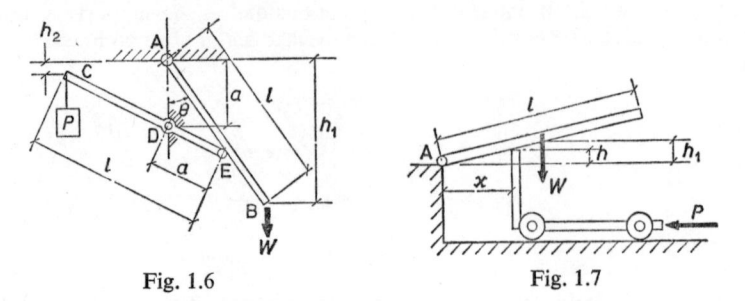

Fig. 1.6   Fig. 1.7

Solution. From the geometry of the figure, triangle ADE is isosceles and angle ADC is equal to 2θ. The heights of the mass centres h_1 and h_2 can be expressed in terms of θ,

$$h_1 = l \cos \theta, \quad h_2 = a - (l-a) \cos 2\theta.$$

If we change θ by a virtual amount $\delta\theta$ the corresponding changes in h_1 and h_2 are obtained by differentiation as in the previous example,

$$\delta h_1 = -l \sin \theta \, \delta\theta, \quad \delta h_2 = 2(l-a) \sin 2\theta \, \delta\theta.$$

The increment of work done during the virtual displacement can be written

$$\delta W = W \delta h_1 + P \delta h_2 = 0.$$

Substituting for δh_1 and δh_2 gives

$$W l \sin \theta \, \delta\theta = 2P(l-a) \sin 2\theta \, \delta\theta.$$

Therefore
$$W = \frac{2P\left(1 - \dfrac{a}{l}\right) \sin 2\theta}{\sin \theta} = 4P\left(1 - \frac{a}{l}\right) \cos \theta.$$

From this we get

$$\theta = \cos^{-1}\left\{\frac{W}{4P\left(1 - \dfrac{a}{l}\right)}\right\}.$$

Example 1.3

The beam shown in Fig. 1.7 is raised by pushing the trolley towards the hinge at A. Estimate the force P required at any distance x between the trolley upright and the support.

VIRTUAL WORK

Solution. If a virtual displacement δx is given to the trolley then, for equilibrium

$$\delta W = 0 = -P\delta x - W\,\delta h_1. \tag{i}$$

We must now find a relationship between h_1 and x. Note that the distance from A to the contact point between the beam and trolley measured along the line of the beam is $\sqrt{x^2+h^2}$. Using proportion therefore we can see from the figure that

$$\frac{h}{\sqrt{x^2+h^2}} = \frac{h_1}{\tfrac{1}{2}l}, \quad h_1 = \frac{l}{2\sqrt{x^2+h^2}}.$$

Differentiating gives

$$\delta h_1 = -\frac{hlx}{2(x^2+h^2)^{\frac{3}{2}}}\,\delta x.$$

Substituting for δh_1 in (i) gives

$$\delta W = 0 = -P\delta x + \frac{Whlx}{2(x^2+h^2)^{\frac{3}{2}}}\,\delta x.$$

Hence

$$P = \frac{Whlx}{2(x^2+h^2)^{\frac{3}{2}}}.$$

Problems

Problem 1.1 Determine the equilibrium position of the system shown in Fig. 1.8 by specifying the height h_1 of the beam A. The cords supporting the beams are perfectly flexible and of length l.

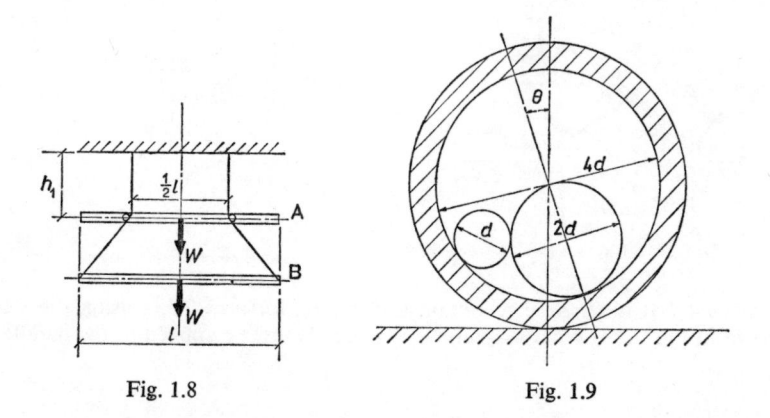

Fig. 1.8

Fig. 1.9

Problem 1.2 Two solid cylinders of equal length and diameters d and $2d$ lie at rest inside a hollow cylinder of inside diameter $4d$ as shown in Fig. 1.9. Determine the angle θ shown in the figure for equilibrium.

14 VIRTUAL WORK AND ENERGY CONCEPTS

Problem 1.3 In the two member system shown in Fig. 1.10, the weight W of each member acts through its mid-point. Determine the moment M required at the support pin A for equilibrium in the position shown in terms of θ (the angle between member AB and the vertical), l and W.

Problem 1.4 In the previous problem, if the moment M is removed and replaced by a horizontal force P on the roller at C, determine the required value of P for equilibrium.

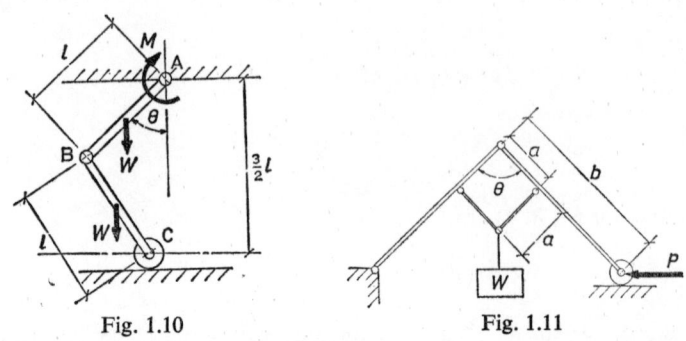

Fig. 1.10

Fig. 1.11

Problem 1.5 Evaluate the equilibrium force P for the mechanism shown in Fig. 1.11 in terms of the member lengths a and b, and the load W. For what ratio of a to b will the mechanism be in equilibrium with no force P applied?

Problem 1.6 In the mechanism shown in Fig. 1.12, the members AB and CBD may be considered weightless. Determine the vertical force required at D to hold the system in the position shown.

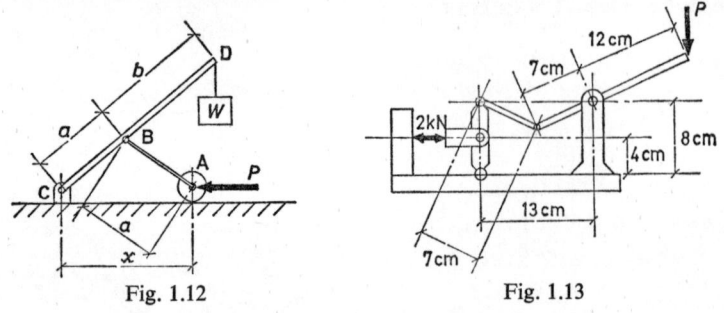

Fig. 1.12

Fig. 1.13

Problem 1.7 It is required to obtain a clamping force of 2 kN using the vice shown in Fig. 1.13. What magnitude of force P must be applied to the handle?

2
STRAIN ENERGY

2.1 Introduction

In the last chapter we examined the behaviour of structures whose members were assumed to be perfectly rigid. In reality structural members are never perfectly rigid, and structures deform whenever forces are applied to them. During deformation the forces are moving against the internal resistance of the body, so that deformation is achieved by work being done on the body.

If the deformations remain when the forces are removed, the material of the body is said to be *plastic* in nature and the work done on the body by the forces has been used up in producing permanent deformation.

If the deformations disappear on removal of the forces and the body resumes its original form, then it is said to be *elastic* in nature, and the work done on the body has been stored as latent energy or *strain energy* which is released on removal of the forces.

The ability of elastic bodies to store energy when deformed has been used by men for many centuries. A prime example of this is the bow and arrow. When an archer pulls the bowstring the bow flexes and stores energy. On release of the bowstring the strain energy stored is released to the arrow in the form of *kinetic energy* and the arrow is propelled away from the bow at high velocity. The storage of energy in this way is also used for less dangerous purposes, clocks and watches operate using the energy stored in their mainsprings. In the field of sport, gymnasts are able to attain great heights jumping from a trampoline which converts the kinetic energy possessed by a gymnast on his descent into strain energy which is released to him to aid his subsequent ascent. In recent years pole vaulters have benefited from the introduction of flexible poles which can be made to store energy at the start of the vault and release this energy when the athlete is in mid-air, thus giving him greater lift.

In engineering the association of strain energy with deformation has very important applications. Engineering structures are generally made from materials which deform very slightly under load. Nevertheless, they do deform and thus store strain energy because of their elasticity.

In this chapter we shall investigate bodies in which no permanent deformations take place and extend the principle of virtual work to deal with elastic structures.

2.2 Virtual work for an elastic body

(a) *Work and complementary work*

Consider the spring system shown in Fig. 2.1 loaded by a force P which gradually increases in magnitude. As the force increases the load point

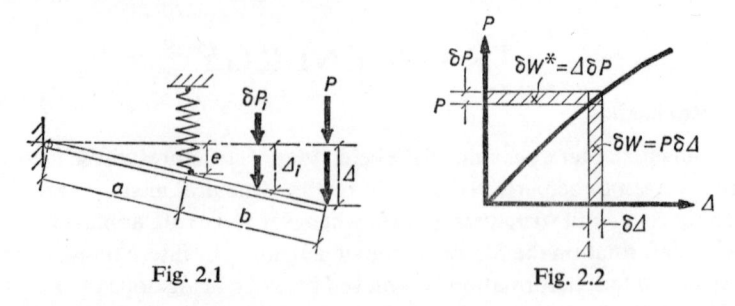

Fig. 2.1

Fig. 2.2

displacement Δ also increases, a general form for the relationship being shown in Fig. 2.2. Since the load varies with deflection, we must take this into account in evaluation of the work done. The work done during any small increase in deflection $\delta\Delta$ is obtained from

$$\delta W = P\delta\Delta. \qquad (2.1)$$

This equality is exact if $\delta\Delta$ is a virtual displacement assumed to take place at constant force. If $\delta\Delta$ is simply an increase in displacement due to an increase in load δP, the work done is given by the area of the shaded vertical strip in Fig. 2.2, $\delta W = P\delta\Delta + \frac{1}{2}\delta P\delta\Delta$. However, if $\delta\Delta$ and δP are infinitesimally small, the term $\frac{1}{2}\delta P\delta\Delta$ becomes negligible and the total work done W by a force P on an elastic body can be obtained from the integral

$$W = \int_0^\Delta P \mathrm{d}\Delta. \qquad (2.2)$$

Therefore the total work done is given by the area under the $P-\Delta$ equilibrium curve.

The area to the left of the $P-\Delta$ curve, which we denote by the symbol W^* is termed *complementary work* and may be obtained by summing the elemental strips of area $\Delta\delta P$ as shown in Fig. 2.2.

$$\delta W^* = \Delta\delta P \qquad (2.3)$$

and

$$W^* = \int_0^P \Delta \mathrm{d}P. \qquad (2.4)$$

This work quantity is rather hypothetical in nature and its physical meaning is not easy to grasp. However, the concepts of complementary work and complementary virtual work are extremely important and form

STRAIN ENERGY

the basis of many energy theorems. One way in which physical significance can be attached to the concept of complementary work is to use the hypothesis that the datum level for load application is the undeflected surface of the structure. This can be explained by re-examining Fig. 2.1. Let us suppose that an arbitrary *virtual load* δP_i is applied at point i on the deflected structure. If this load is applied at the undeflected level of point i, then it must move through a distance Δ_i before it meets the deflected structure. Thus the load has done an amount of work equal to $\delta P_i \Delta_i$ which is lost and has no effect on the structure. But $\delta P_i \Delta_i$ is the complementary work done by the load, and we can therefore think of complementary work as the work lost by the load before it reaches the deflected structure. Equation (2.3) holds whether δP is a virtual force or an incremental addition to the applied force. In the latter case the total complementary work can be obtained by integrating to give equation (2.4).

Note that we now have two distinct virtual work forms for an elastic body; virtual work as a product of $P\delta\Delta$, i.e. a real force moving through a virtual displacement, and virtual complementary work as a product of $\Delta\delta P$, i.e. a virtual force moving through a real displacement.

Note also that the effects of a number of forces acting on a body simultaneously can be taken into account by adding the effects of each force, for example, if a number of forces $P_1, P_2, \ldots, P_n$ act simultaneously the total work done is

$$W = \sum_{i=1}^{i=n} \int_0^{\Delta_i} P_i \mathrm{d}\Delta_i.$$

(b) *Strain energy and complementary strain energy*

In a loaded body internal forces and deformations are present. Just as the externally applied forces do work on the body due to the displacements, internal work is done by the displacements of the internal forces. In an elastic body this internal work is stored as strain energy. Let us examine once more the spring system of Fig. 2.1. The spring is displaced by an amount e and the force in the spring is F, as shown in Fig. 2.3. Now if any increase in external loading occurs, the spring displacement will increase by an amount δe and the spring force by an amount δF. Thus the spring itself can be considered as an elastic body under load. However, the spring is an internal part of the structure. The spring load and deformation are internal effects and as such are not considered in the evaluation of the external work done. Instead these are used to evaluate the internal work which is stored as strain energy. The strain energy u and complementary

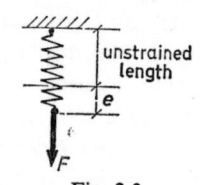

Fig. 2.3

strain energy u^* are obtained in the same way as the external work and complementary work giving

$$\delta u = F\,\delta e, \quad u = \int_0^e F\,de \tag{2.5}$$

$$\delta u^* = e\,\delta F, \quad u^* = \int_0^F e\,dF. \tag{2.6}$$

Many springs, particularly of the coil type illustrated are *linearly elastic*, i.e. their displacements are proportional to the internal forces. In such a case we have

$$F = Ke, \tag{2.7}$$

where K is the spring stiffness. For this type of spring

$$\delta u = Ke\,\delta e, \quad u = \int_0^e Ke\,de = \tfrac{1}{2}Ke^2 \tag{2.8}$$

$$\delta u^* = \frac{F}{K}\,\delta F, \quad u^* = \int_0^F \frac{F}{K}\,dF = \tfrac{1}{2}\frac{F^2}{K}. \tag{2.9}$$

Note that since $F = Ke$, equations (2.8) and (2.9) are identical, in other words the strain energy stored in a linear elastic body is equal to the complementary strain energy.

To obtain energy expressions for a body which is completely elastic, such as a beam, we must examine the behaviour of an infinitesimally small element of the body as shown in Fig. 2.4. When discussing the actions and

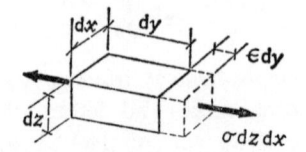

Fig. 2.4

deformations present on such an element, we use the terms *stress* (force per unit area) and *strain* (deformation per unit original length) rather than force and displacement. In Fig. 2.4, a direct stress σ is acting on the element, causing a strain ε. The total force on the element is $\sigma\,dx\,dz$ and the extension is $\varepsilon\,dy$. We can think of the element as an infinitesimally small spring with the given force and deformation, and hence obtain expressions for the energy stored in the element by substitution in equations (2.5) and (2.6),

$$\delta u_e = \sigma\,dx\,dz \times \delta\varepsilon\,dy = \sigma\,\delta\varepsilon\,dV \tag{2.10}$$

$$\delta u_e^* = \varepsilon\,dy \times \delta\sigma\,dx\,dz = \varepsilon\,\delta\sigma\,dV, \tag{2.11}$$

where $dV\,(=dx\,dy\,dz)$ is the volume of the element.

STRAIN ENERGY

The virtual strain energy in the complete body is found by integrating throughout the body,

$$\delta u = \int_{VOL} \sigma \, \delta\varepsilon \, dV \qquad (2.12)$$

$$\delta u^* = \int_{VOL} \varepsilon \, \delta\sigma \, dV. \qquad (2.13)$$

In the case of a body subjected to shear stresses, we use τ the shear stress and γ the shear strain in place of σ and ε.

To obtain the total strain energy corresponding to given stress and strain conditions, we must integrate the equations (2.12) and (2.13) to obtain expressions analogous to (2.5) and (2.6). In the particular case of linear elastic bodies, we have

$$\sigma = E\varepsilon, \qquad (2.14)$$

where E is Young's Modulus of Elasticity, the spring constant of the material. Substituting into equations (2.10) and (2.11) we obtain

$$\delta u_e = E\varepsilon \, \delta\varepsilon \, dV.$$

Therefore

$$u_e = \int_0^\varepsilon E\varepsilon \, d\varepsilon \, dV = \tfrac{1}{2} E\varepsilon^2 \, dV. \qquad (2.15)$$

Similarly

$$u_e^* = \int \frac{\sigma}{E} \, d\sigma \, dV = \tfrac{1}{2} \frac{\sigma^2}{E} \, dV. \qquad (2.16)$$

Therefore

$$u = \int_{VOL} \tfrac{1}{2} E\varepsilon^2 \, dV. \qquad (2.17)$$

and

$$u^* = \int_{VOL} \tfrac{1}{2} \frac{\sigma^2}{E} \, dV. \qquad (2.18)$$

Since $\sigma^2 = E^2 \varepsilon^2$, then $u = u^*$ for linear elastic structures. Note that when dealing with virtual work and virtual changes in strain energy, we do not require to evaluate the total strain energy since we use equations (2.12) and (2.13) in analysis. Some energy methods, however, such as Castigliano's methods which we shall encounter later, use the total strain energy expressions. This is the reason for their derivation in this section.

(c) *Virtual work for an elastic body*

Since we are dealing with perfectly elastic bodies, it follows from the principle of conservation of energy that the work done by the external forces on a body is equal to the internal energy gained by the body. This equality holds for all infinitesimal force changes and is also true for

20 VIRTUAL WORK AND ENERGY CONCEPTS

complementary work. For an elastic body the principle of virtual work states, therefore

$$\delta W = \delta u. \tag{2.19}$$

To illustrate the application of this equation we shall examine the system of Fig. 2.1.

Example 2.1

Given that the spring in Fig. 2.1 has stiffness K and is unloaded when the rigid bar is horizontal, determine the force P required to displace the bar an amount Δ under the load.

Solution From the geometry of the system the spring extension e can be expressed in terms of the displacement Δ.

$$e = \Delta \times \frac{a}{a+b}.$$

So that if the system is given a virtual displacement $\delta\Delta$ the spring extension is

$$\delta e = \delta\Delta \times \frac{a}{a+b}.$$

We therefore have $\Delta W = P\,\delta\Delta$, and from equation (2.8),

$$\delta u = Ke\,\delta e = K\Delta\,\frac{a}{a+b} \times \delta\Delta\,\frac{a}{a+b}$$

since the spring is linear elastic.

The principle of virtual work states that

$$P\,\delta\Delta = K\Delta\,\delta\Delta \times \frac{a^2}{(a+b)^2}.$$

Therefore

$$P = \frac{Ka^2}{(a+b)^2}\,\Delta.$$

This example was very simple. The application of virtual work to more complex problems involving linear springs is illustrated by the following two examples

Example 2.2

In the arrangement shown in Fig. 2.5 the bar ABC is rigid. The spring is linear of stiffness K and stretches from the pin at A to the slider at B. It has its neutral position when the bar is horizontal. Obtain the value of P for equilibrium at any angle θ.

Solution If we apply a virtual rotation $\delta\theta$ to the bar from its equilibrium position then from the principle of virtual work we have $\delta W = \delta u$, where

$$\delta W = P\,\delta h \quad \text{and} \quad \delta u = Ke\,\delta e.$$

STRAIN ENERGY

Therefore

$$P\,\delta h = Ke\,\delta e. \qquad \text{(i)}$$

From the geometry of the system $h = l \sin \theta$. We obtain δh in terms of $\delta\theta$ by differentiation of this expression.

$$\delta h = l \cos \theta \, \delta\theta.$$

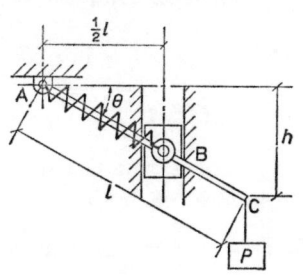

Fig. 2.5

Also from the geometry of the system, the length of the spring at angle θ is

$$\frac{l}{2 \cos \theta}.$$

Since the spring has its neutral position when $\theta = 0$, then its original length is $\tfrac{1}{2}l$. The extension of the spring e is therefore $e = \tfrac{1}{2}l(\sec \theta - 1)$.

δe is found by differentiation

$$\delta e = \tfrac{1}{2}l \sec \theta \tan \theta \, \delta\theta.$$

Substituting in (i) gives

$$Pl \cos \theta \, \delta\theta = K\tfrac{1}{2}l(\sec \theta - 1) \times \tfrac{1}{2}l \sec \theta \tan \theta \, \delta\theta.$$

From this we obtain

$$P = \tfrac{1}{4}Kl \sec^2 \theta \tan \theta \,(\sec \theta - 1).$$

Example 2.3

The two link member shown in Fig. 2.6 has a torsion spring at joint B. The spring is in its neutral position (i.e. non-deformed) when both links are upright. Express the force P for equilibrium in terms of the spring stiffness K, the length of each link $\tfrac{1}{2}l$, and the angle α from the vertical of each link.

Solution The deformation associated with a torsion spring is a rotation, $e = \theta$, since θ is the relative rotation of the links. The strain energy contained in such a spring is therefore $u = \tfrac{1}{2}K\theta^2$. Now if we displace the load point by an amount δh, we must ensure the equality

$$P(-\delta h) = \delta u = K\theta \, \delta\theta. \qquad \text{(i)}$$

The minus sign before δh is included, since in this case a negative δh adds work to the body.

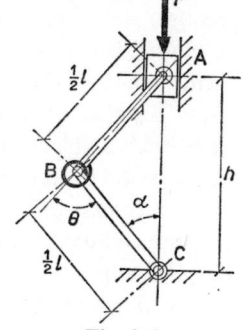

Fig. 2.6

Now from the geometry of the body we have
$$h = l\cos\alpha \quad \text{and} \quad \theta = 2\alpha.$$

We can therefore find the virtual displacements of h and θ in terms of α by differentiation,
$$\delta h = -l\sin\alpha\,\delta\alpha \quad \text{and} \quad \delta\theta = 2\delta\alpha.$$
Substituting in (i) gives
$$Pl\sin\alpha\,\delta\alpha = K \times 2\alpha \times 2\delta\alpha$$
$$P = \frac{4K\alpha}{l\sin\alpha}.$$

An interesting point to be noted from this solution is that as $\alpha \to 0$ then $\alpha \simeq \sin\alpha$ so that when α is zero, i.e. both links are vertical, the solution obtained gives $P = \frac{4K}{l}$. Thus if P is less than $\frac{4K}{l}$ the structure remains straight and only deflects when P reaches this value. This value of P is called the critical buckling load, or instability load for the structure.

Additional problems are provided at the end of the chapter.

2.3 Systems with many degrees of freedom

In all problems encountered so far, the virtual displacements of any part of the structure under consideration have been uniquely determined by the applied virtual displacement. For instance, in example 2.3 the displacements $\delta\alpha$, $\delta\theta$ and δh are completely interdependent. In these cases, therefore, if we postulate one virtual displacement we can immediately obtain the virtual displacement system throughout the structure. This type of structure is said to have *one degree of freedom*.

Many structures have more than one degree of freedom. For example, the structure of Fig. 2.7 (a) is able to displace independently in two ways, i.e. by change in θ_1 and by change in θ_2, so that we require to specify both angles to determine the position of any part of the structure. Thus it has two degrees of freedom. The structure of Fig. 2.7 (b) has five degrees of freedom, since the angles of four members and the position of one joint are required to specify its position. The elastic beam of Fig. 2.7 (c) has an infinite number of degrees of freedom since its deformed position has an infinite number of variations.

For such a structure we can apply any virtual displacement system we choose to aid us in our analysis, so long as the virtual displacement system satisfies *compatibility*, i.e. no breaks or discontinuities occur in the structure, joints do not come apart, etc. and the boundary restraints are complied with.

We can use this arbitrariness of choice to obtain solutions to problems involving such structures with the minimum of labour. For example, if a

STRAIN ENERGY 23

structure has a number of undefined loads acting on it and we wish to find the magnitude of one of these loads, we choose a virtual displacement system due to which only that load moves. Thus the virtual external work done is dependent only on the magnitude of the particular load we require

(a)

(b)

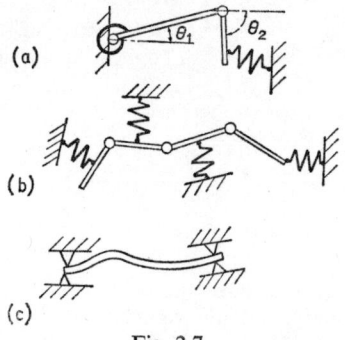

(c)

Fig. 2.7

to find, and equating this to the virtual change in strain energy gives the magnitude of this load immediately. This method is illustrated in the next example.

Example 2.4

In the structure shown in Fig. 2.8 linear springs of stiffness K extend from the centre points of links AB and BC. The top ends of the springs can move horizontally so that the springs remain vertical at all times. If the springs are in their

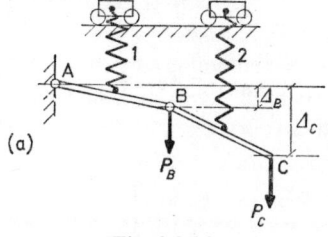

(a)

Fig. 2.8 (a)

neutral positions when both links are horizontal, obtain the forces P_B and P_C in terms of Δ_B and Δ_C.

Solution To find P_C we hold link AB rigid and displace the end C of BC to give a virtual displacement of $\delta\Delta_C$ vertically to point C. The work done is

$$\delta W = P_C \delta \Delta_C.$$

The strain energy increase is solely due to the extension of spring 2, since spring 1 does not extend. The total extension of spring 2 from its neutral position is $\Delta_B + \frac{1}{2}(\Delta_C - \Delta_B)$. Thus $e_2 = \frac{1}{2}(\Delta_B + \Delta_C)$. Now

$$F_2 = Ke_2 = \frac{1}{2}K(\Delta_B + \Delta_C).$$

During the virtual displacement $\delta\Delta_C$ the mid-point of BC displaces by $\frac{1}{2}\delta\Delta_C$, so that we obtain

Therefore
$$P_C\delta\Delta_C = \tfrac{1}{2}K(\Delta_B+\Delta_C)\times\tfrac{1}{2}\delta\Delta_C.$$

$$P_C = \tfrac{1}{4}K(\Delta_B+\Delta_C).$$

To find P_B we cannot apply a displacement system such that P_B moves with P_C remaining stationary. However, we can apply such a system that P_C moves only

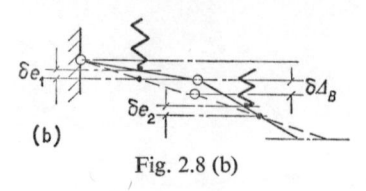

Fig. 2.8 (b)

horizontally, as shown above, and thus does no work. The virtual displacements of springs 1 and 2 are, from the geometry of the displacement, both equal to $\tfrac{1}{2}\delta\Delta_B$. The total extension of spring 2 is as before and that of spring 1 is $\tfrac{1}{2}\Delta_B$. We have therefore

This gives
$$P_B\delta\Delta_B = \tfrac{1}{2}K(\Delta_B+\Delta_C)\tfrac{1}{2}\delta\Delta_B + K\tfrac{1}{2}\Delta_B\tfrac{1}{2}\delta\Delta_B.$$

$$P_B = \tfrac{1}{4}K(2\Delta_B+\Delta_C).$$

Note that if we had applied two different arbitrary virtual displacements both of which allowed P_B and P_C to do work, then we could have obtained two simultaneous equations in P_B and P_C the solution of which would have furnished the answers obtained here. The solution of simultaneous equations is not required if we choose our virtual displacement systems so that only one force does work.

A point which should be remembered with regard to systems with many degrees of freedom is that, if the actual deflected form of the system is specified completely, then this specification must be based on the assumption that the number of unknown forces acting on the system (excluding reactions) is equal to the number of degrees of freedom. For example, two forces P_B and P_C were assumed for the structure of the previous problem. For a system with only one degree of freedom one force is sufficient to specify the displaced form. In a continuous beam if the displacements are prescribed at all points along the beam then in general forces will be distributed at all points.

In the following example we have a two degree of freedom system with only one unspecified load. We therefore have to obtain the displaced form before evaluating the load.

Example 2.5

In the system shown in Fig. 2.9 a load P is applied horizontally at the end of bar ABC and displaces this end horizontally by 0.5 metres. Determine the

STRAIN ENERGY

magnitude of P given that spring 1 has stiffness of 0.25 kN/m and spring 2 stiffness of 1 kN/m and are in their neutral positions when both bars are vertical.

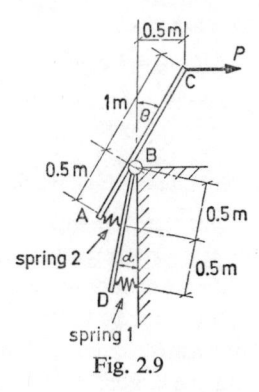

Fig. 2.9

Solution Here we can evaluate θ immediately

$$\theta = \sin^{-1}\frac{0.5}{1} = 30°.$$

The load P however, depends upon both θ and α, so that we must evaluate α before we can obtain P.

From the figure we can obtain expressions for the extension of springs 1 and 2.

$$e_1 = 2 \sin \tfrac{1}{2}\alpha \text{ m} \qquad \text{(i)}$$

$$e_2 = 2 \times 0.5 \sin \tfrac{1}{2}(\theta - \alpha) \text{ m}. \qquad \text{(ii)}$$

Now we shall apply a virtual rotation $\delta\alpha$ to bar BD keeping θ constant. The virtual spring extensions are

$$\delta e_1 = 1 \cos \tfrac{1}{2}\alpha \, \delta\alpha$$

$$\delta e_2 = -\tfrac{1}{2} \cos \tfrac{1}{2}(\theta - \alpha) \, \delta\alpha.$$

Since no external work is done during the virtual displacement, we have

$$\delta W = \delta u = K e_1 \delta e_1 + K e_2 \delta e_2 = 0,$$

$$0.25 \times 2 \sin \tfrac{1}{2}\alpha \cos \tfrac{1}{2}\alpha \, \delta\alpha - \sin \tfrac{1}{2}(\theta - \alpha) \times \tfrac{1}{2} \cos \tfrac{1}{2}(\theta - \alpha) \, \delta\alpha = 0.$$

Since $\sin A \cos A = \tfrac{1}{2} \sin 2A$, this becomes

$$\tfrac{1}{4} \sin \alpha - \tfrac{1}{4} \sin (\theta - \alpha) = 0.$$

Hence $\alpha = \theta - \alpha$, $\therefore \alpha = \tfrac{1}{2}\theta$. Thus for any value of θ we can find α. In this case $\theta = 30°$, therefore $\alpha = 15°$. Substituting in equations (i) and (ii) gives

$$e_1 \simeq 0.261 \text{ m}, \quad e_2 \simeq 0.130 \text{ m}.$$

To obtain P we now apply a virtual displacement $\delta\theta$ with α remaining constant giving

$$\delta e_1 = 0, \quad \delta e_2 = \tfrac{1}{2}\delta\theta \cos 7.5° = 0.483 \, \delta\theta.$$

26 VIRTUAL WORK AND ENERGY CONCEPTS

The horizontal distance moved by P during the virtual displacement is

$$\delta h = \cos\theta \; \delta\theta.$$

Applying the virtual work equation gives

$$P \; \delta h = Ke_2 \; \delta e_2$$

$$P \times 0.866 \; \delta\theta = 1 \times 0.130 \times 0.483 \; \delta\theta.$$

Hence $P \simeq 0.073$ kN.

2.4 The principle of virtual complementary work

So far we have dealt with problems in which virtual work has been used to obtain equilibrium loads for structures which are displaced in a given fashion. Many problems which are encountered in the analysis of structures, however, are concerned either directly or indirectly with the evaluation of the deflections of structures whose loads are known. This type of problem can be tackled directly using the *principle of virtual complementary work*.

The expressions for complementary work and complementary strain energy were derived in §2 of this chapter. The principle of virtual complementary work simply states that

$$\delta W^* = \delta u^*. \tag{2.20}$$

Since $\delta W^* = \Delta \delta P$, we can see that using this principle we apply *virtual forces* δP to a body, evaluate the change in complementary strain energy, and hence obtain the real displacements Δ. This method is illustrated in the next example.

Example 2.6

Here we have the structure of example 2.4. This time we are given loads P_B and P_C and require to find the deflections Δ_B and Δ_C in terms of the loads.

Solution In using the principle of complementary work we must apply the equations of equilibrium to the structure, both for real loadings and for virtual

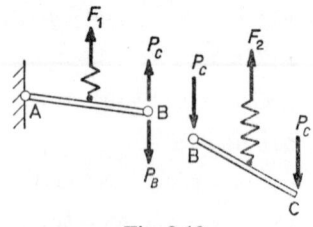

Fig. 2.10

loads. First we evaluate F_1 and F_2. To do this we split the structure into parts AB and BC as shown.

Taking moments about B for BC gives $F_2 = 2P_C$, and also equilibrium of vertical forces for BC gives a downward force equal to P_C at B. This applies an upward force of the same magnitude in member AB.

STRAIN ENERGY

Taking moments about A for AB gives $F_1 = 2(P_B - P_C)$.

To find Δ_C we apply a virtual load δP_C at C and evaluate the virtual spring forces δF_1 and δF_2 in the same way as we did for the real forces, obtaining

$$\delta F_1 = 2\delta P_C, \quad \delta F_2 = -2\delta P_C.$$

From $\delta W^* = \delta u^*$ we obtain

$$\Delta_C \delta P_C = \frac{F_1 \delta F_1}{K} + \frac{F_2 \delta F_2}{K}$$

$$= \frac{1}{K}[2P_C \times 2\delta P_C + 2(P_B - P_C) \times (-2\delta P_C)].$$

Hence
$$\Delta_C = \frac{4}{K}(2P_C - P_B).$$

To evaluate Δ_B we apply a virtual load δP_B at B and proceed as for Δ_C.
This time
$$\delta F_1 = 2\delta P_C, \quad \delta F_2 = 0.$$

Therefore
$$\Delta_B \delta P_C = 2(P_B - P_C) \times \frac{2\delta P_C}{K},$$

and hence
$$\Delta_B = \frac{4}{K}(P_B - P_C).$$

2.5 The unit load method

As we have seen, the principle of virtual complementary work is used to obtain the real displacements in loaded structures through the application of virtual loads. The general statement of the principle may be expressed

$$\Sigma \delta P . \Delta = \int_{VOL} \varepsilon \, \delta\sigma \, dV. \tag{2.21}$$

In the case of structures whose elasticity is obtained from springs, the right-hand side of this equation can be replaced by $\Sigma e \delta F$.

The left-hand side of the equation gives the external virtual complementary work in terms of virtual loads and real displacements. The right-hand side gives the internal virtual complementary work, or virtual strain energy, in terms of virtual internal forces and the real internal deformations. The internal virtual forces must be in equilibrium with the applied external virtual forces. In the general case if a structure is very flexible, such as is the case with most of the systems we have examined, the geometry of the structure varies with loading and we cannot obtain an exact analysis of the equilibrium situation unless we know how it is displaced. Since we are trying to find the displacements, then we obviously do not know what they are, and so for this type of problem we must resort to trial and error methods or iterative methods of approach, which can be very long winded and time consuming.

28   VIRTUAL WORK AND ENERGY CONCEPTS

Happily the majority of engineering structures deform only very slightly, and for a great many problems we can assume that displacements are so small that the geometry of the structure does not change. In addition, very many structures are linear elastic within their working range. Under these conditions problems become very much simpler to handle, since we can relate the real internal deformations to the real internal forces, and hence to the real external forces. We can therefore obtain the real internal deformations by virtue of an equilibrium analysis of the structure under the real loading conditions. To find the virtual internal forces we can remove all real loads from the structure and apply the virtual forces to the structure alone.

Now, if we wish to obtain the deflection of a given point in the structure we simply apply a single virtual force at that point in the direction of the required displacement. Because of the arbitrariness of the magnitude of the virtual force we can set this equal to unity. Equation (2.21) therefore becomes

$$\Delta_i \times 1 = \int_{VOL} \varepsilon\, \delta\sigma\, dV \quad (\text{or } \Delta_i \times 1 = \Sigma e\, \delta F). \tag{2.22}$$

From this equation, called the *unit load equation*, we can obtain the required displacement immediately; $\delta\sigma$ (or δF) is obviously the virtual internal force system due to the unit virtual load.

The use of this equation is illustrated by the next example.

Example 2.7

In the cantilever type system shown in Fig. 2.11 evaluate the displacement at the tip and at the centre joint C on the assumption that the displacements are small.

Solution First we find the actual moments on each spring F_A, F_B, F_C and F_D by taking moments about each spring in turn, considering moments to the right-hand side only,

$$F_D = Pl, \quad F_C = 2Pl, \quad F_B = 3Pl, \quad F_A = 4Pl.$$

Note that although the actions of the torsion springs are moment actions we still use the symbol F to denote these actions for convenience. For this type of spring the associated displacement e is a rotation and has no units, so that the product $F \times e$ has units of work as for linear springs. Now we remove the force P and

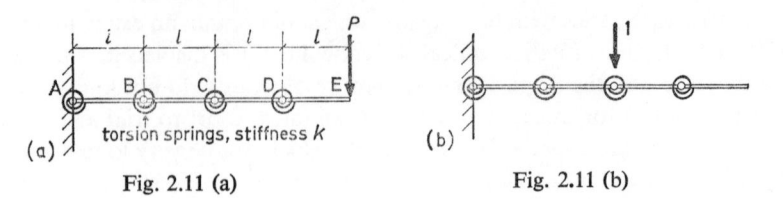

(a) torsion springs, stiffness k

(b)

Fig. 2.11 (a)   Fig. 2.11 (b)

STRAIN ENERGY 29

apply a virtual force of unity at the tip to obtain the tip deflections. The virtual spring moments may be evaluated directly by proportion from the real case

$$\delta F_D = l, \quad \delta F_C = 2l, \quad \delta F_B = 3l, \quad \delta F_A = 4l.$$

The principle of virtual complementary work is now applied in the form of the unit load equation giving

$$\Delta_E \times 1 = \Sigma e \, \delta F = \Sigma \frac{F \, \delta F}{K}$$

$$\frac{1}{K}(Pl \times l + 2Pl \times 2l + 3Pl \times 3l + 4Pl \times 4l) = 30 \frac{Pl^2}{K}.$$

To obtain Δ_C we apply a virtual load of unity, or a *unit load*, at C as shown, giving

In this case
$$\delta F_D = 0, \quad \delta F_C = 0, \quad \delta F_B = l, \quad \delta F_A = 2l.$$

$$\Delta_C \times 1 = \frac{1}{K}(3Pl \times l + 4Pl \times 2l) = 11 \frac{Pl^2}{K}.$$

The equivalent method following from the principle of virtual work is the *unit displacement method*, whereby virtual displacements are set arbitrarily equal to unity for the sake of convenience.

2.6 Castigliano's theorem

One of the longest standing and most widely used methods of structural analysis follows from this theorem, originally developed by Alberto Castigliano (1847-1884), an Italian engineer. This theorem follows from consideration of the total strain energy of a system and has two parts, one following from the equality of total external work and internal strain energy and the other following from the equality of external complementary work and internal complementary strain energy. To derive Part 1 of this theorem, we recall $W = \Sigma \int_0^{\Delta_i} P_i \, d\Delta_i$. Since $W = u$, we can write

$$u = \Sigma \int_0^{\Delta_i} P_i \, d\Delta_i$$

$$= \int_0^{\Delta_1} P_1 \, d\Delta_1 + \int_0^{\Delta_2} P_2 \, d\Delta_2 + \ldots + \int_0^{\Delta_i} P_i \, d\Delta_i + \ldots \int_0^{\Delta_n} P_n \, d\Delta_n.$$

Differentiating both sides of this equation with respect to Δ_i we obtain

$$\frac{\partial u}{\partial \Delta_i} = P_i. \tag{2.23}$$

This is Castigliano's Theorem, Part 1. To derive Part 2 we use complementary work and complementary strain energy and obtain

$$u^* = W^* = \Sigma \int_0^{P_i} \Delta_i \, dP_i = \int_0^{P_1} \Delta_1 \, dP_1 + \ldots + \int_0^{P_i} \Delta_i \, dP_i + \ldots + \int_0^{P_n} \Delta_n \, dP_n.$$

30 VIRTUAL WORK AND ENERGY CONCEPTS

Differentiating with respect to P_i we find

$$\frac{\partial u^*}{\partial P_i} = \Delta_i. \tag{2.24}$$

This is Castigliano's Theorem, Part 2.

Two examples are now used to show the application of Castigliano's Theorem.

Example 2.8

Repeat Example 2.4 using Castigliano's Theorem, Part 1.

Solution From the solution of Example 2.4 we have

$$e_1 = \tfrac{1}{2}\Delta_B \quad \text{and} \quad e_2 = \tfrac{1}{2}(\Delta_B + \Delta_C).$$

Using Castigliano's method we obtain the strain energy u in terms of Δ_B and Δ_C.

$$u = \tfrac{1}{2}Ke_1^2 + \tfrac{1}{2}Ke_2^2 = \tfrac{1}{8}K\Delta_B^2 + \tfrac{1}{8}K(\Delta_B+\Delta_C)^2.$$

To obtain P_C we differentiate u with respect to Δ_C,

$$P_C = \frac{\partial u}{\partial \Delta_C} = \tfrac{1}{4}K(\Delta_B+\Delta_C), \text{ as before. Similarly,}$$

$$P_B = \frac{\partial u}{\partial \Delta_B} = \tfrac{1}{4}K\Delta_B + \tfrac{1}{4}K(\Delta_B+\Delta_C) = \tfrac{1}{4}K(2\Delta_B+\Delta_C), \text{ as before.}$$

Example 2.9

Evaluate the tip deflection of the structure of Example 2.7 using Castigliano's Theorem, Part 2.

Solution From Example 2.7 we know the actual spring moments

$$F_A = 4Pl, \quad F_B = 3Pl, \quad F_C = 2Pl, \quad F_D = Pl.$$

The strain energy contained in the structure is found from $u = \Sigma\tfrac{1}{2}\dfrac{F^2}{K}$.

$$u = \frac{1}{2K}[(4Pl)^2 + (3Pl)^2 + (2Pl)^2 + (Pl)^2] = 15\frac{P^2l^2}{K}.$$

Castigliano's Theorem, Part 2 states $\dfrac{\partial u^*}{\partial P_E} = \Delta_E$

and for a linear elastic system $u^* = u$.

Applying this we obtain $\dfrac{\partial u^*}{\partial P} = \Delta_E = 30\dfrac{Pl^2}{K}$ as before.

In Example 2.7 we also evaluated the displacement of the centre joint. But in this case, since we have no force at the centre joint, we cannot

STRAIN ENERGY 31

obtain the central deflection. Therefore we can see a limitation inherent in this method, i.e. we cannot obtain displacements at an unloaded point of the structure. There are ruses which we can employ to overcome this and evaluate the displacement of unloaded points using Castigliano's method. We shall not examine these since the unit load method can be used to obtain displacements of any point in a structure whether loaded or unloaded.

Castigliano's Theorem offers an elegant method of analysis of structures. It has decreased in popularity in recent years, however, due to inherent limitations, such as that mentioned, in comparison with the virtual work methods, and for this reason we shall use the other methods, in particular the unit load method, in future analyses.

Problems

Problem 2.1 In the system shown in Fig. 2.12, the lever swings through an arc of 45° each side of the centre line. Given that the unstrained length of the spring is $\frac{3}{2}R$ evaluate the force P required to keep the lever in position at any angle θ.

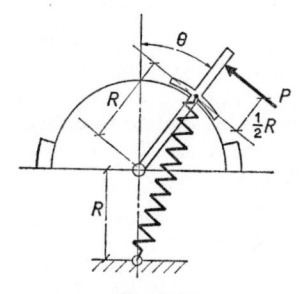

Fig. 2.12

Problem 2.2 The springs in the system shown in Fig. 2.13 can take either tensile or compressive forces. Determine the angle θ of the bars for equilibrium if the unstrained length of all springs is $\frac{1}{2}l$.

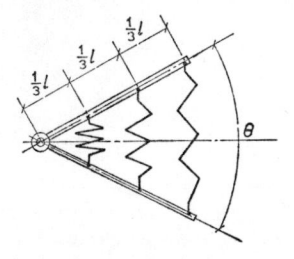

Fig. 2.13

32 VIRTUAL WORK AND ENERGY CONCEPTS

Problem 2.3 The torsion spring on the system shown in Fig. 2.14 has its neutral position when the tube is horizontal. Its stiffness is 200 N/radian. The rope carrying the load *P* passes freely through the tube. Determine the load *P* required to make the tube lie at an angle of 45° as shown.

Fig. 2.14

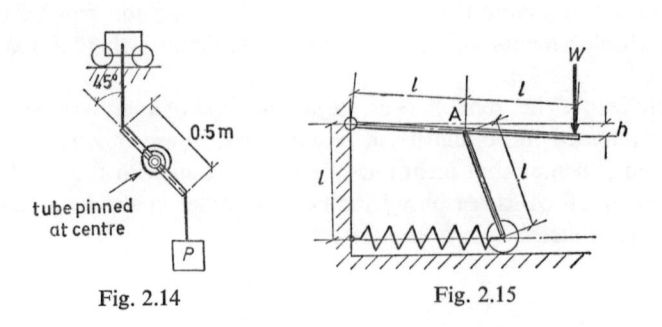

Fig. 2.15

Problem 2.4 In the structure shown in Fig. 2.15 the stiffness of the spring is K and its unstrained length is l. Determine the equilibrium position of the structure by specifying h for a given W. Assume that h is small enough to make horizontal movement of joint A negligible.

Problem 2.5 The top end of the rigid bar ABCD in Fig. 2.16 is displaced 100 mm due to a horizontal load *P*. Noting that no force acts on the bottom end, determine the horizontal displacement of this end and the magnitude of force *P*. *K* for each spring is 2 kN/m.

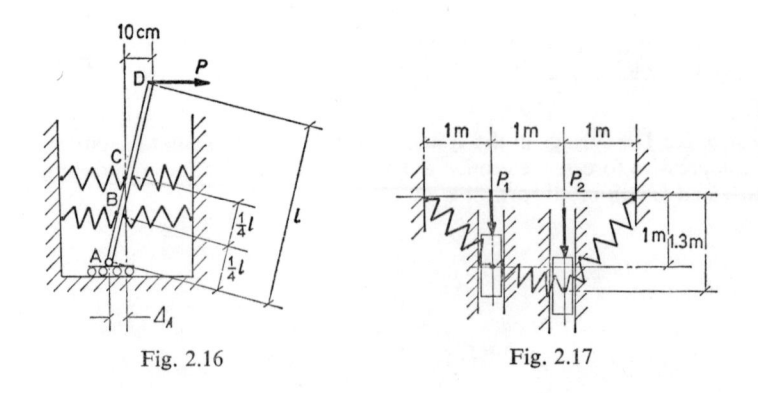

Fig. 2.16

Fig. 2.17

Problem 2.6 Determine the forces P_1 and P_2 applied to the pistons in the system shown in Fig. 2.17. All springs are of stiffness 5 kN/m and have unstrained length of 1 metre.

STRAIN ENERGY 33

Problem 2.7 The structure shown in Fig. 2.18 consists of two bars each of length $\frac{1}{2}l$ joined by a torsion spring of stiffness K_1 with one bar joined also to a base by a torsion spring of stiffness K_2. If both springs are undeformed when both bars are upright, obtain an expression for the vertical load P required to make the end of the top bar touch the base as shown. Find also the angle θ shown in the figure.

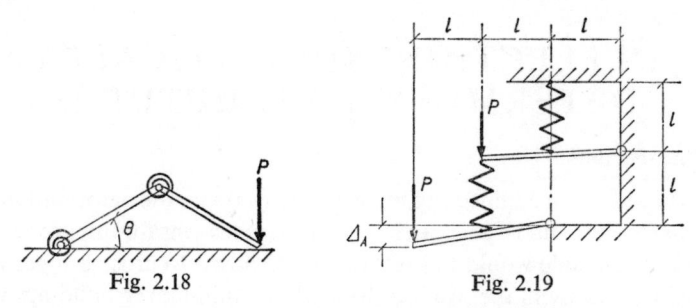

Fig. 2.18   Fig. 2.19

Problem 2.8 Evaluate the displacement Δ_A in terms of P, l and the stiffness K of both springs for the system shown in Fig. 2.19 using the assumption that displacements are small. Attempt solutions from both unit load and Castigliano approaches.

C

3

DEFLECTIONS OF STATICALLY DETERMINATE STRUCTURES

3.1 Introduction

It is necessary in engineering to be able to evaluate the deformations in a body or structure under load. In many cases the reasons for this are obvious, for example people would not feel readily disposed to cross bridges which swayed noticeably in the wind, or to work in multi-storey buildings which displayed the same characteristic. On a smaller scale, in the design of machinery care must be taken to ensure that distortion and twisting of component parts is not so great as seriously to impair the efficiency of a machine, e.g. if shafts are allowed to deflect overmuch then gears attached to them may not mesh properly.

Another very important reason for the necessity of deformation analyses of structures, which is not quite so obvious, is that in many structures the stresses under load cannot be determined if the deformation characteristics are not known. This type of structure is known as a *statically indeterminate structure*. Since the safe and efficient design of structures depends upon a detailed knowledge of the stress situation, then it is of prime importance that engineers should be equipped with the means of analysing the deformation characteristics of structures.

In this chapter and the next we shall confine our examination to *statically determinate structures*, i.e. structures in which loads and reactions can be obtained directly from the equations of equilibrium. From this description it is apparent that we shall be dealing with known loads in most cases and using these to evaluate displacements. In view of this the energy principles which are most useful are generally those which follow from the principle of virtual complementary work. Also we shall use the limitations of *small deflections* throughout, since engineering structures generally deflect only slightly. As a further simplification we shall only consider *linear elastic* materials, since this is a practical situation for many engineering structures.

We shall use the *unit load equation* for the analysis of all deflection problems. To use this equation we must be able to obtain expressions for virtual changes in complementary energy under various loading conditions.

STATICALLY DETERMINATE STRUCTURES

The next section in this chapter is therefore devoted to obtaining expressions for virtual complementary energy in terms of direct force, bending moment, shear force and torque, starting from the basic equation.

3.2 Formulation of virtual complementary energy expressions for various loading conditions

For a linear elastic material we know that $\varepsilon = \sigma/E$. The virtual complementary energy can therefore be written as $\delta u^* = \int_{VOL} \sigma \dfrac{\delta\sigma}{E} \, dV$. We now use this expression to obtain δu^* in terms of applied loadings.

(a) Direct force

The elemental length of bar shown in Fig. 3.1 is loaded by a force P. The line of action of P passes through the centroid of the bar section so that the stress σ produced by the load is constant over the complete section and is of magnitude $\sigma = P/A$.

If a virtual force of magnitude p is transmitted to the bar the virtual stress $\delta\sigma$ is similarly given by $\delta\sigma = p/A$. Therefore

$$\delta u^* = \int_{VOL} \frac{\sigma \, \delta\sigma}{E} \, dV = \int_{VOL} \frac{Pp}{EA^2} \, dV. \tag{3.1}$$

Since $dV = dA \, dx$, this becomes, on integration over the cross-sectional area,

$$\delta u^* = \int_0^l \frac{Pp}{AE} \, dx, \tag{3.2}$$

where l is the total length of the bar.

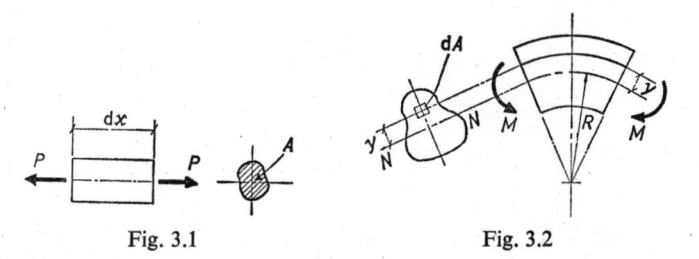

Fig. 3.1   Fig. 3.2

(b) Simple bending

Fig. 3.2 shows an elemental length of a beam bent by a moment into an arc of a circle of radius R to the centroid of the section. The beam section is symmetrical about the *neutral axis* of the section, which passes through the centroid of the section so that simple bending theory is applicable.

The equations of simple bending theory are

$$\frac{\sigma}{y} = \frac{M}{I} = \frac{E}{R}, \qquad (3.3)$$

where I is a geometric property termed the *second moment of area* of the section about its neutral axis and is found from the integral $I = \int_A y^2 \, dA$.

The stress σ can therefore be put in terms of M, i.e. $\sigma = My/I$. Similarly the virtual stress $\delta\sigma$ due to a virtual moment m can be written $\delta\sigma = my/I$. Substituting for σ and $\delta\sigma$ gives therefore

$$\delta u^* = \int_{VOL} \frac{\sigma \, \delta\sigma}{E} \, dV = \int_{VOL} \frac{Mm}{EI^2} y^2 \, dV = \int_0^l \frac{Mm}{EI^2} \left[\int_A y^2 \, dA \right] dx. \qquad (3.4)$$

Since $\int_A y^2 \, dA = I$, we have

$$\delta u^* = \int_0^l \frac{Mm}{EI} \, dx. \qquad (3.5)$$

(c) *Shear force*

Shear force acts on a beam if the bending moment varies along the beam. For general loading on a beam, therefore, shear force and bending moment appear simultaneously. In general it is found that the effects of shear are very small in comparison to those of bending moment, except in the case of very short beams, and the contribution of shear effects is usually neglected. The virtual complementary strain energy of shear forces is therefore derived here only for the sake of completeness and in most problems we tackle it will be neglected.

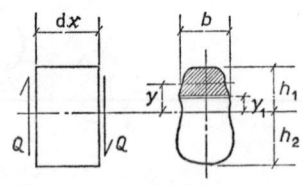

Fig. 3.3

The relationship between the shear force Q and the shear stress on an elementary strip parallel to the neutral axis of the element cross section, as shown in Fig. 3.3, is given by the formula

$$\tau = \frac{Q\bar{A}}{Ib}, \qquad (3.6)$$

STATICALLY DETERMINATE STRUCTURES

where $\bar{A}$ is the first moment of area of the section above y_1, about the centroid (or the first moment of the area below y_1, if y_1 is negative).

$$\bar{A} = \int_{y_1}^{h_1} y \, \mathrm{d}A. \tag{3.7}$$

If a virtual shear force q acts on the elementary strip, the virtual shear stress is similarly obtained from $\delta\tau = q\bar{A}/Ib$. Recalling that the shear stresses τ and $\delta\tau$ are treated the same as direct stresses for the evaluation of δu^*, and also that the elastic shear modulus is denoted by G, we have

$$\delta u^* = \int_{VOL} \frac{\tau \, \delta\tau}{G} \, \mathrm{d}V = \int_{VOL} \frac{Qq\bar{A}^2}{I^2 G b^2} \, \mathrm{d}V = \int_{VOL} \frac{Qq}{GI^2 b^2} \left[\int_{y_1}^{h_1} y \, \mathrm{d}A \right]^2 \mathrm{d}V. \tag{3.8}$$

This can be rewritten as

$$\delta u^* = \int_0^l \frac{qQ}{GI^2} \left[\int_{h_2}^{h_1} \left(\int_{h_2}^{h_1} y \, \mathrm{d}A \right)^2 \frac{\mathrm{d}y_1}{b} \right] \mathrm{d}x. \tag{3.9}$$

This expression is very unwieldy and requires undue labour in use. To avoid this we can first assume that as an approximation the shear stresses are constant over the cross-section, i.e. $\tau = Q/A$, $\delta\tau = q/A$. In these circumstances the virtual complementary energy can easily be found in an analogous manner to that for direct stress, giving

$$\delta u^* = \int_0^l \frac{Qq}{GA} \, \mathrm{d}x. \tag{3.10}$$

This function does not give the same value for δu^* as the accurate expression (3.9), but it can be made to yield identical results simply by using it in conjunction with a multiplying factor K.

$$\delta u^* = K \int_0^l \frac{Qq}{GA} \, \mathrm{d}x, \tag{3.11}$$

where K is a function of the cross-sectional geometry and is obtained by equating expressions (3.11) and (3.9). K is tabulated for many common engineering shapes in structural reference books.

(d) *Torsion*

The stresses and deformations produced by the action of a twisting moment or torque T on an elemental length of circular section beam as shown in Fig. 3.4, are given by the expression

$$\frac{\tau}{r} = \frac{T}{J} = G \frac{\mathrm{d}\theta}{\mathrm{d}x}, \tag{3.12}$$

where J is a geometrical property known as the *torsion constant*.

For a circular section $J = 2I = \int r^2 \, dA$. For the annular strip shown in Fig. 3.4, $\tau = Tr/J$. Similarly, if a virtual torque t is applied $\delta\tau = tr/J$.
Therefore

$$\delta u^* = \int_{VOL} \frac{\tau \, \delta\tau}{G} \, dV = \int_{VOL} \frac{Ttr^2}{GJ^2} \, dV. \quad (3.13)$$

Since $dV = dx \, dA$ this can be written

$$du^* = \int_0^l \frac{Tt}{GJ^2} \left[\int r^2 \, dA \right] dx. \quad (3.14)$$

Noting the definition of J for a circular section, we arrive at

$$\delta u^* = \int_0^l \frac{Tt}{GJ} \, dx. \quad (3.16)$$

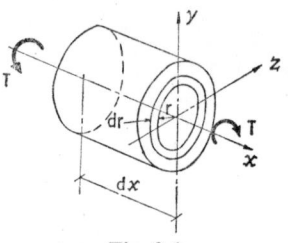

Fig. 3.4

The analysis of beams of non-circular cross-section is outside the scope of this book since a detailed knowledge of torsion behaviour is required in order to evaluate the torsion constant J. For an examination of the torsion of non-circular sections the reader is referred to A. C. Walker, *Torsion*, in this series.

(e) *Combined effects*

If all four types of actions are present on a structure, the total virtual complementary strain energy is found by adding that due to each action.

$$\delta u^* = \int_0^l \frac{Pp}{AE} \, dx + \int_0^l \frac{Mm}{EI} \, dx + K \int_0^l \frac{Qq}{GA} \, dx + \int_0^l \frac{Tt}{GJ} \, dx. \quad (3.16)$$

3.3 Deflections of pin-jointed frames

Many engineering structures are built up of a number of straight members connected at their ends to form a framework. The connections are usually welded or riveted in practice, but as a first approximation in the analysis of this type of structure they are assumed to be pinned so that moments cannot be transmitted from one member to another. Using this assumption, the work involved in analysis is greatly reduced. In many

STATICALLY DETERMINATE STRUCTURES

cases the moments occurring in the members have little effect on the deflections of a frame, since these are governed mainly by the axial shortening of the members. If the moments arising from the rigidity of the joints are required these can be estimated using as a basis the results of the pinned frame analysis.

A typical frame structure is shown in Fig. 3.5. To ensure that axial loads only are present in the members all forces are assumed to act at the pin joints. The external supports generally encountered are of the roller type as at point A or the pin type as at B.

The axial forces in the bars can be evaluated using various methods. If we give a virtual displacement to any of the joints of the frame of Fig. 3.5, we obtain conclusions similar to those obtained from the analysis of the virtual work conditions for a particle, i.e. the joint must be in equilibrium under the actions of all forces acting on it, including external loads, member forces and reaction forces.

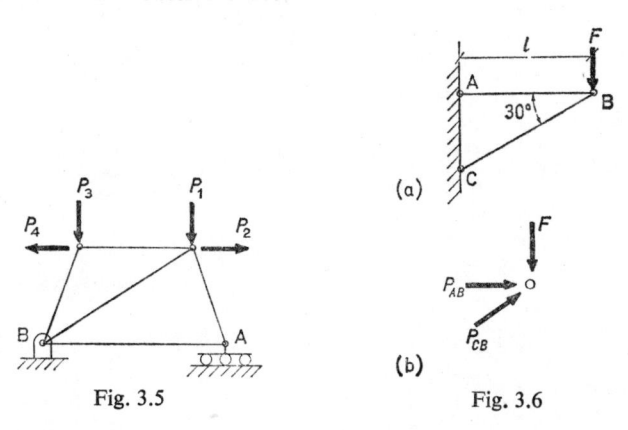

Fig. 3.5

Fig. 3.6

The method used in evaluating the deflections are illustrated by the following examples:

Example 3.1

Determine the vertical deflection of the load point of the two-bar frame shown in Fig. 3.6 in terms of the load F, the cross-sectional area of the bars A, the length of the horizontal bar l, and Young's modulus for the material.

Solution First we must evaluate the forces in the members. We do this by considering the equilibrium of joint B. At this joint the forces present are as shown. The force P_{CB} in bar CB is found from equilibrium of vertical forces,

$$P_{CB} \sin 30° = F$$

$$P_{CB} = \frac{F}{\sin 30°} = 2F.$$

VIRTUAL WORK AND ENERGY CONCEPTS

Equilibrium of horizontal forces gives

$$P_{AB} - P_{CB} \cos 30° = 0$$

and hence

$$P_{AB} = -\sqrt{3}F \simeq -1.732\,F.$$

The negatives sign for P_{AB} indicates that the force acts in the opposite direction from that assumed, i.e. a tensile force acts in bar AB.

To obtain the deflection using the unit load method, we apply a unit load vertically at B. The bar forces, P_{AB} and p_{BC}, can be determined in this case by proportion from the actual loading,

$$P_{AB} = -1.732, \quad p_{BC} = 2.$$

The virtual change in complementary energy is

$$\delta u^* = \frac{P_{AB}p_{AB}}{AE} l + \frac{P_{BC}p_{BC}}{AE} \frac{l}{\cos 30}.$$

Therefore

$$1 \times \Delta_B = \delta u^* = \frac{3Fl}{AE} + \frac{8Fl}{\sqrt{3}AE} = 7.61 \frac{Fl}{AE}.$$

Example 3.2

In the frame structure shown in Fig. 3.7 all members are of cross-sectional area 200 mm². Young's modulus for the material is $E = 200$ kN/mm². Determine the vertical deflection of point C.

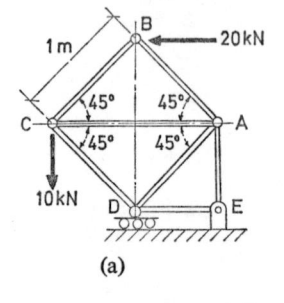

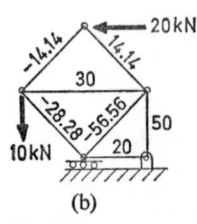

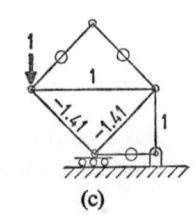

Fig. 3.7

Solution First we evaluate the forces in each member. To do this we consider the equilibrium of each joint in turn. For joint B: equilibrium of vertical forces gives $P_{CB} = -P_{BA}$, equilibrium of horizontal forces gives

$$(P_{CB} - P_{BA}) \frac{1}{\sqrt{2}} = -20 \text{ kN}.$$

Hence

$$2P_{BA} = 20\sqrt{2} \text{ kN}.$$

Therefore

$$P_{BA} = 14.14 \text{ kN (tensile)}, \quad P_{BC} = -14.14 \text{ kN (compressive)}.$$

STATICALLY DETERMINATE STRUCTURES

We now examine joint C and apply the equilibrium equations at this joint, noting that P_{CB} is compressive as it was at joint B. Equilibrium of vertical forces gives

$$\frac{\sqrt{2}}{2} P_{CD} = -10 - 14.14 \frac{\sqrt{2}}{2}.$$

Hence
$$P_{CD} = -28.28 \text{ kN}.$$

Equilibrium of horizontal forces gives

$$P_{CA} = 14.14 \frac{\sqrt{2}}{2} + 28.28 \frac{\sqrt{2}}{2} = 30 \text{ kN}.$$

We now move to joints A and D, applying the equilibrium equations to find the forces in all bars. These are shown in Fig. 3.7 (b).

In order to determine the deflection at any point on the frame we apply a unit load at the point in the direction of the deflection required and evaluate the bar forces due to the unit load. To find the vertical deflection of C we apply a unit load vertically at C. The forces in each bar are determined in the same way as the real forces and are shown in Fig. 3.7 (c).

The unit displacement equation states that

$$\Delta_A = \sum \frac{Ppl}{AE}.$$

Since there are a number of members it is advisable to perform the summation in a tabular form as shown below.

Member	P kN	p	$\frac{L}{AE}$ mm/kN	$\frac{PpL}{AE}$ mm
AB	14.14	0	0.025	0
BC	−14.14	0	0.025	0
CD	−28.28	−1.41	0.025	1
DA	−56.56	−1.41	0.025	2
AE	50	1	0.0177	0.685
DE	20	0	0.0177	0
AC	30	1	0.0354	1.06

$$\Sigma = 4.745 \text{ mm}$$

Therefore $\Delta_C = 4.745$ mm.

This routine procedure is applicable for the evaluation of deflections in any pin-jointed frame.

PROBLEMS

Problem 3.1 In the structure of Example 3.2, the bars AC, AE and AD are removed and replaced by bars 400 mm² cross-sectional area. Under these conditions determine the horizontal deflection of joint B under the given loading.

42 VIRTUAL WORK AND ENERGY CONCEPTS

Problem 3.2 Determine the horizontal and vertical deflections of joint A on the frame shown in Fig. 3.8. Area of all bars = 300 mm². $E = 207$ kN/mm².

Fig. 3.8

Fig. 3.9

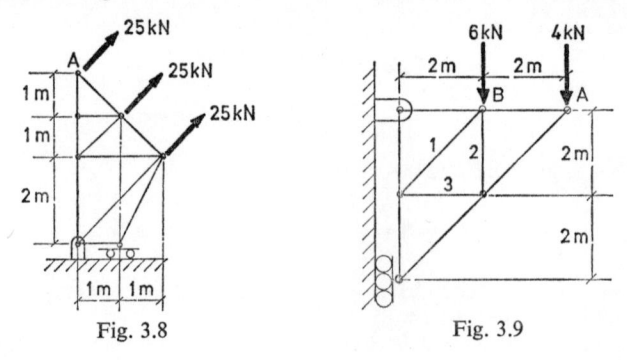

Problem 3.3 In the frame shown in Fig. 3.9 the stress in all members is 100 N/mm². Evaluate the vertical deflections of both load points. $E = 200$ kN/mm².

Problem 3.4 In the structure of the previous example, it is required to decrease the deflection of point B by increasing the area of the members labelled 1, 2 and 3. What is the least deflection obtainable at B by this method? What effect does this have on the deflection of point A.

Problem 3.5 For the frame shown in Fig. 3.10 determine the resultant deflection of the load point in magnitude and direction. Area of all bars is 1000 mm². $E = 200$ kN/mm².

Fig. 3.10

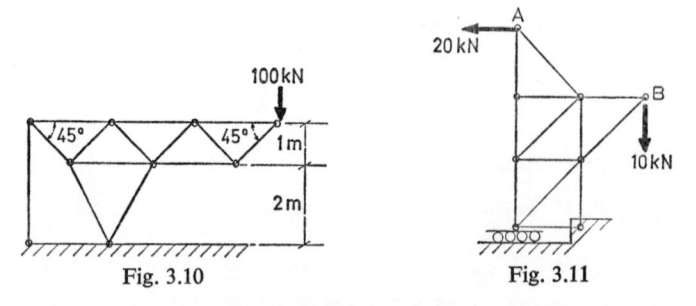

Fig. 3.11

Problem 3.6 In the frame shown in Fig. 3.11 all horizontal and vertical members are 1 m long. Determine the resultant deflections of joints A and B. Using these results evaluate the change in distance between these two points. Area of bars = 500 mm². $E = 200$ kN/mm².

Problem 3.7 For the structure of the previous example, deduce the physical meaning of the deflection obtained if two unit loads are applied at A and B in the direction of the line AB, acting towards B and A respectively. Evaluate this deflection when the structure is under the same loads as in Fig. 3.11. Compare the result with the results of the previous example.

4

DEFLECTIONS OF BEAMS

4.1 Introduction

In this chapter we apply the *unit load equation* to the analysis of statically determinate beam systems. In this case all four load actions, direct force, bending moment, torque and shear force may be present. In general, the effects of shear force on deflection are negligible in comparison with those of bending. For example, it can be shown that for a rectangular section cantilever of depth d and length l, the tip deflection produced by a load at the tip is proportional to l/d due to shear and is proportional to $(l/d)^3$ due to bending. For a very short deep beam with depth equal to length (i.e. $l/d = 1$) the ratio of shear deflection to bending deflection is $\frac{3}{4}$. For a longer beam ($l/d = 10$) the ratio of shear deflection to bending deflection is reduced to 3/400. Since most beams used in practice are long in comparison to their depth, we can see that shear effects may be neglected with little loss in accuracy. In view of this we shall not consider shear effects in analyses to determine deflections.

The support reactions generally encountered on the analysis of beams are roller supports, or simple supports which prevent displacement in one direction, pinned ends which prevent displacement in any direction, and fixed or built-in supports which prevent displacement and also prevent change in slope of the beam at the supported point.

4.2 Deflections of beams

The application of the unit load equation to the evaluation beam deflections is best shown by illustrative examples.

Example 4.1

Evaluate the central deflection of the simply supported beam shown in Fig. 4.1 loaded by w per unit length.

Solution We first obtain the reactions at both supports and evaluate the bending moment at any point on the beam by summing moments to the left of that point. For this beam $R_A = R_B = \frac{1}{2}wl$ and the moment at any point distance x from support A is $R_A x - \frac{1}{2}wx^2$. (Note that anticlockwise moments to the left of the point are given a negative sign.) Since $R_A = \frac{1}{2}wl$ the bending moment M at any point is

$$M = \tfrac{1}{2}wlx - \tfrac{1}{2}wx^2.$$

We now apply a unit load at the centre point, evaluate the reactions and obtain the bending moment M at any point. This is

$$m = \tfrac{1}{2}x \text{ for } x < \tfrac{1}{2}l, \quad m = \tfrac{1}{2}l - \tfrac{1}{2}x \text{ for } x > \tfrac{1}{2}l.$$

The M and m diagrams are shown in Fig. 4.1 (a) and (b) respectively. Since the beam and loading are symmetrical about the centre line, we see that the complementary energy of the complete beam is twice that of one symmetrical half of the beam. Applying the unit load equation, we have

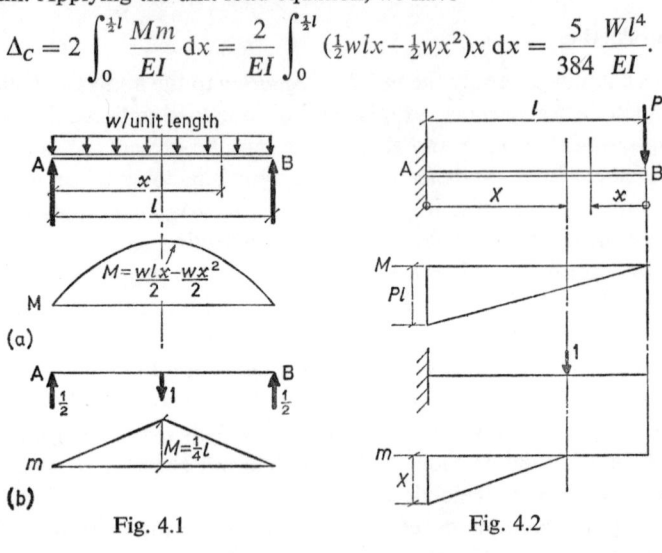

$$\Delta_C = 2\int_0^{\frac{1}{2}l} \frac{Mm}{EI} \, dx = \frac{2}{EI}\int_0^{\frac{1}{2}l} (\tfrac{1}{2}wlx - \tfrac{1}{2}wx^2)x \, dx = \frac{5}{384}\frac{Wl^4}{EI}.$$

Fig. 4.1

Fig. 4.2

Example 4.2

Obtain an expression for the deflection at any point distance X from the fixed end of a tip loaded cantilever.

Solution The cantilever shown in Fig. 4.2 has a vertical tip load. The deflection at point C distance X from A is found by applying a unit load here. The bending moment diagrams M and m are shown in Fig. 4.2. At any point, $M = px$. For $x < l - X$, $m = 0$. For $x > l - x$, $m = x - l + X$.
The required displacement is

$$\Delta_C = \int_0^l \frac{Mm}{EI} \, dx = \int_0^X 0 \, dx + \int_{l-x}^l \frac{Px(x - l + X)}{EI} \, dx.$$

Integrating gives

$$\Delta_C = \frac{P}{EI}[\tfrac{1}{3}x^3 - \tfrac{1}{2}x^2(l - X)]_{l-x}^l = \frac{Pl^3}{EI}\left[\tfrac{1}{2}\left(\frac{X}{l}\right)^2 - \tfrac{1}{6}\left(\frac{X}{l}\right)^3\right].$$

Evaluation of the slope or twist at a point

If we wish to find the slope at a point in a beam we apply a unit moment at that point. Similarly we can find the twist at a point by applying a unit torque. The use of a unit moment application is shown in the next example.

DEFLECTIONS OF BEAMS

Example 4.3

The beam ABC shown in Fig. 4.3 is loaded by a distributed load of 2 kN/m over part AB. The second moment of area for BC is 10^6 mm^4 and that for AB is 3×10^6 mm^4. Evaluate the vertical deflection of B and the slope at C. $E = 200$ kN/mm^2.

Solution First we obtain the reactions R_A and R_C and draw the M diagram. We then apply a unit load at B to find Δ_B and a unit moment at C to find θ_C. The bending moment diagrams for each system are shown in Fig. 4.3.
  Now

$$\Delta_B = \int \frac{Mm_1}{EI} dx = \frac{1}{EI_{AB}} \int_0^2 (3x - x^2)\tfrac{1}{2}x \, dx + \frac{1}{EI_{BC}} \int_0^2 (2-x)(1-\tfrac{1}{2}x) \, dx$$

$$= \frac{2}{EI_{AB}} + \frac{4}{3EI_{BC}} \text{ kN m}^3$$

(note the change of origin for x).
Substituting for EI and bringing all dimensions to mm gives

$$\Delta_B = \frac{2 \times 10^9}{200 \times 300 \times 10^4} + \frac{4 \times 10^9}{3 \times 200 \times 100 \times 10^4} = 10 \text{ mm}.$$

Now

$$\theta_C = \int \frac{Mm_2}{EI} dx = \frac{1}{EI_{AB}} \int_0^2 (3x - x^2)\tfrac{1}{4}x \, dx + \frac{1}{EI_{BC}} \int_0^2 (2-x)(\tfrac{1}{2} + \tfrac{1}{4}x) \, dx$$

$$= \frac{1}{EI_{AB}} + \frac{4}{3EI_{BC}} \text{ kN m}^2 = 0.008 \, 33 \text{ radians}$$

$$= 0.476°.$$

Fig. 4.3

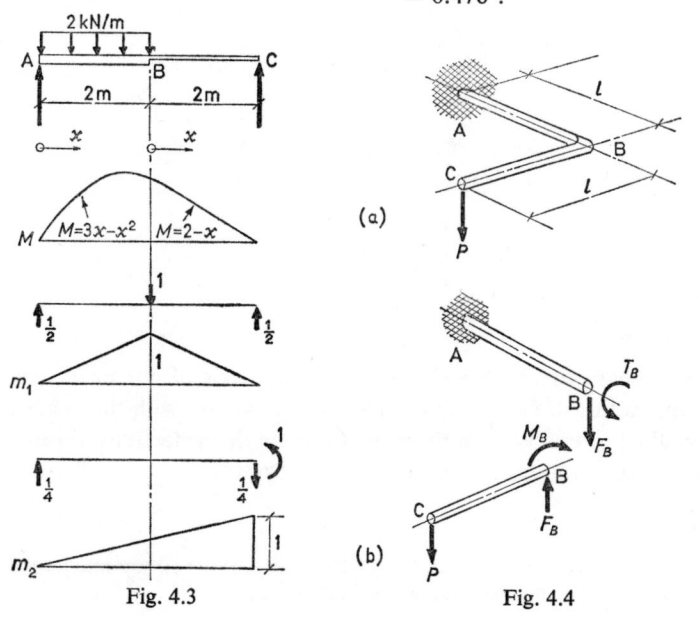

Fig. 4.4

46 VIRTUAL WORK AND ENERGY CONCEPTS

So far we have considered only beam bending. In the next example the effects of torsion are illustrated.

Example 4.4

The cranked beam ABC shown in Fig. 4.4 is of circular cross-section and $J = 2I$. Obtain an expression for the deflection of point C due to a load P applied at C. Assume that $G = 0.4E$.

Solution In order to examine this beam we split it into two parts, AB and BC. We must apply the conditions of equilibrium to each part individually. Since we split the beam at B we must assume that moments and shear forces act on each bar at this point.

Considering the equilibrium of part BC we have $F_B = P$ and $M_B = Pl$. Now the force F_B on part BC must be counteracted by an equal and opposite force F_B on AB. The moment M_B is transformed into a torque T_B on AB of equal magnitude. Thus on part AB we have $M = Px$ and on part BC we have $M = Px$ and $T = Pl$.

If we apply a unit vertical load at C we obtain the forces at B in the same way. These are proportional to the real forces in this case since the virtual loading is proportional to the real loading. Therefore on AB, $m = x$ and on BC, $m = x$ and $t = l$.

Now

$$\Delta_C = \int \frac{Mm}{EI} dx + \int \frac{Tt}{GJ} dx,$$

where the integrals extend over both parts

$$= \int_0^l \frac{Px.x}{EI} dx + \int_0^l \frac{Px.x}{EI} dx + \int_0^l \frac{Pl.l}{GJ} dx$$

$$= \frac{Pl^3}{3EI} + \frac{Pl^3}{3EI} + \frac{Pl^3}{GJ} = Pl^3 \left[\frac{2}{3EI} + \frac{1}{GJ} \right].$$

Substituting $G = 0.4E$ and $J = 2I$ gives

$$\Delta_C = Pl^3 \left[\frac{2}{3EI} + \frac{1}{0.8EI} \right] = \frac{23}{12} \frac{Pl^3}{EI}.$$

Curved beams

Simple beam theory is suitable for the analysis of curved beams provided the radius of curvature is large in comparison with the beam cross-section dimensions. We can therefore evaluate the deflections of thin rings, etc., using the same methods as for straight beams.

Example 4.5

Evaluate the horizontal deflection of the load point of the semicircular ring shown in Fig. 4.5. Consider bending effects only.

Fig. 4.5

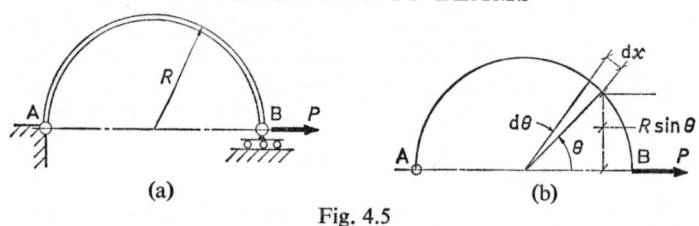

Solution We first find the moment at any point due to the applied loading. (Note that this is the only force at B since taking moments about A shows that the vertical reaction is zero.) The moment caused by P about a section at an angle θ from the horizontal is seen from the sketch to be $M_\theta = Pl\sin\theta$. Applying a unit horizontal force at B gives $m_\theta = 1R\sin\theta$. Thus

$$\Delta_B = \int \frac{Mm}{EI}\,dx = \int_0^\pi \frac{Pl\sin\theta \times l\sin\theta}{EI} R\,d\theta.$$

(Note $dx = R\,d\theta$.)

Integrating gives
$$\Delta_B = \frac{\pi}{2}\frac{PR^3}{EI}.$$

4.3 Structures with several beam and bar members

We have already seen that we can analyse multi-bar frames by splitting them into their component bars and adding the complementary energy (or change in complementary energy) of each to obtain that for the complete frame. This same type of analysis can be used for structures composed of beams or combinations of beams and bars. Some examples are used to demonstrate this.

Example 4.6

The structure shown in Fig. 4.6 is formed from two circular rods AB and CD bent into quadrants of a circle and joined by a beam BC of the same section. Determine the central deflection of the beam in terms of EI and l, if a uniformly distributed load of p per unit length acts from B to C. Assume $GJ = 0.8EI$.

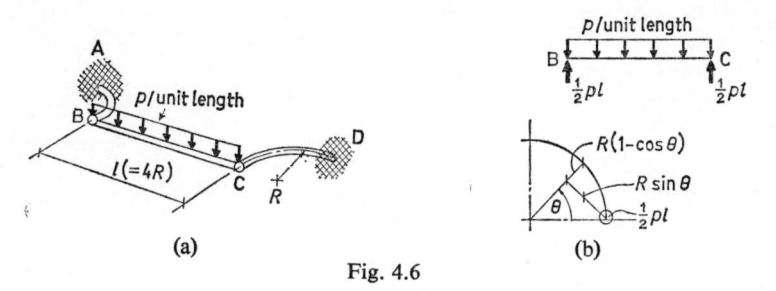

Fig. 4.6

VIRTUAL WORK AND ENERGY CONCEPTS

Solution The beam BC is supported at its ends by reactions from the rods of magnitude $\frac{1}{2}pl$ at each end, thus the beam can be represented as in Fig. 4.6. Each rod is now loaded by the reaction force $\frac{1}{2}pl$ which causes bending and torsion in the rod. The moment and torque at any point on the rods are

$$M_R = \tfrac{1}{2}pl\, R \sin \theta$$

$$T_R = \tfrac{1}{2}pl\, R(1 - \cos \theta).$$

The moment on the beam is

$$M_B = \tfrac{1}{2}plx - \tfrac{1}{2}px^2 = \tfrac{1}{2}p(lx - x^2).$$

Now applying a unit load at the beam centre gives reaction forces on the rods of $\frac{1}{2}$. In this case

$$m_B = \tfrac{1}{2}x \quad \text{for} \quad x < \tfrac{1}{2}l$$

$$m_R = \tfrac{1}{2}R \sin \theta$$

$$t_R = \tfrac{1}{2}R(1 - \cos \theta).$$

Because of a symmetry about the centre point of beam BC the strain energy contained in the structure is twice that contained in one symmetrical half of the structure, thus we can consider only one half of the structure.

$$\Delta_{CENTRE} = 2 \left[\int_0^{\frac{1}{2}l} \frac{M_B m_B}{EI}\, dx + \int_0^{\frac{1}{2}\pi} \frac{M_R m_R}{EI}\, dx + \int_0^{\frac{1}{2}\pi} \frac{T_R t_R}{GJ}\, dx \right].$$

Now

$$\int_0^{\frac{1}{2}l} \frac{M_B m_B}{EI}\, dx = \frac{1}{EI} \int_0^{\frac{1}{2}l} \frac{p}{2}(lx - x^2)\frac{x}{2}\, dx = \frac{5}{768} \frac{pl^4}{EI}$$

and

$$\int_0^{\frac{1}{2}\pi} \frac{M_R m_R}{EI}\, dx = \frac{Pl}{EI} \int_0^{\frac{1}{2}\pi} \tfrac{1}{4}R^3 \sin^2 \theta\, d\theta = \frac{Pl}{EI} \times \frac{\pi R^3}{16} = \frac{\pi}{1\,024} \frac{pl^4}{EI}$$

and

$$\int_0^{\frac{1}{2}\pi} \frac{T_R t_R}{GJ}\, dx = \frac{Pl}{GJ} \int_0^{\frac{1}{2}\pi} \tfrac{1}{4}R^3(1-\cos \theta)^2\, d\theta = \frac{Pl}{GJ} \times \frac{\left(\frac{3\pi}{4} - 2\right)R^3}{4}$$

$$= \frac{\left(\frac{3\pi}{4} - 2\right) Pl^4}{0.8 \times 256 EI}.$$

Therefore

$$\Delta_{CENTRE} = 2 \left[\frac{5}{768} + \frac{\pi}{1\,024} + \frac{3\pi - 8}{819.2} \right] \frac{Pl^4}{EI} \simeq 0.022\,6\, \frac{Pl^4}{EI}.$$

DEFLECTIONS OF BEAMS

Example 4.7

The structure shown in Fig. 4.7 is made up of a beam ABC and pin-jointed bars AD, AE, BD, DE and BE. Determine the horizontal and vertical deflections of C for the load shown. Neglect direct action on ABC. $E = 200$ kN/mm², $I_{ABC} = 5$ cm⁴, area of bars $= 2$ cm².

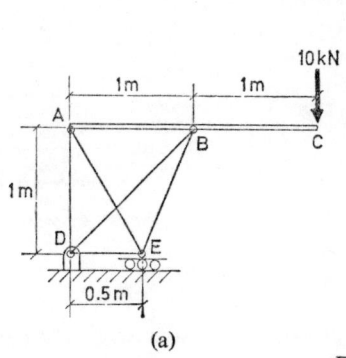

Fig. 4.7

Solution To obtain forces in the component parts of a thin structure we must look at the overall structure and at the individual parts.

Taking moments about D for the complete structure gives

$$\Sigma M_D = 0, \text{ therefore } R_E \times \tfrac{1}{2} = 10 \times 2, \text{ hence } R_E = 40 \text{ kN}\uparrow.$$

Equilibrium of joint E gives $P_{EB} = P_{EA} = \dfrac{40}{\sqrt{3}}$ kN (both compressive). Equilibrium of vertical forces for the complete structure gives $R_D = 30$ kN. Equilibrium of joint D gives $P_{DA} = 30$ kN (tensile)
$$P_{DB} = 0.$$

Using the bar forces we can find the forces on beam ABC.
Resolving forces at A and B gives

$$V_A = 10 \text{ kN}\downarrow$$

$$V_B = 20 \text{ kN}\uparrow$$

$$H_A = 10 \text{ kN}\leftarrow$$

$$H_B = 10 \text{ kN}\rightarrow.$$

The M and m_1 distributions in ABC are shown in Fig. 4.7 (b). Since direct actions are neglected in evaluating the strain energy of ABC, we can neglect the direct

D

loads at A and B. For the bars the loads due to a vertical unit load at C are in proportion to those of the actual load. Thus

$$\Delta_{C\,(vertical)} = \int \frac{Mm}{EI}\,dx + \sum \frac{Ppl}{AE}$$

$$= \frac{2}{200 \times 10^3 \times 5 \times 10^4}\left[\frac{10^6 \times 10^2 \times 66.7}{2}\right]$$

$$+ \left(2 \times \frac{40}{\sqrt{3}} \times 10^3 \times \frac{4}{\sqrt{3}} \times \frac{200}{\sqrt{3}} \times \frac{1}{200 \times 10^3 \times 2 \times 10^2}\right)$$

$$+ \frac{30 \times 10^3 \times 3 \times 100}{200 \times 10^3 \times 2 \times 10^2}$$

$$\simeq 0.87 \text{ mm}.$$

To evaluate $\Delta_{C(horizontal)}$ we apply a unit load at C horizontally. Now ΣM_D gives $R_E = 2$ and $p_{CA} = P_{EB} = \frac{2}{\sqrt{3}}$ (both compression). Equilibrium of vertical and horizontal forces gives $V_D = 2\downarrow$ and $H_D = 1\leftarrow$.

Equilibrium of joint D gives $p_{AD} = 1$, $p_{BD} = \sqrt{2}$ (both tension). Due to this unit load, $m_{ABC} = 0$ (check by evaluating forces at joints A and B). Thus

$$\Delta_{C\,(horizontal)} = \sum \frac{Ppl}{AE}$$

$$= \frac{2 \times 40 \times 10^3}{\sqrt{3}} \times \frac{2}{\sqrt{3}} \times \frac{200}{\sqrt{3}} \times \frac{1}{200 \times 10^3 \times 2 \times 10^2}$$

$$+ \frac{30 \times 10^3 \times 1 \times 100}{200 \times 10^3 \times 2 \times 10^2}$$

$$\simeq 0.13 \text{ mm}.$$

Problems

Problem 4.1 Derive an expression for the central deflection of a uniform beam, simply supported at both ends, under the action of a point load distance X from one end.

Problem 4.2 The cantilever shown in Fig. 4.8 has a second moment of area I from C to B and $2I$ from B to A. Determine the vertical deflection of C in terms of w, E, I and l.

DEFLECTIONS OF BEAMS   51

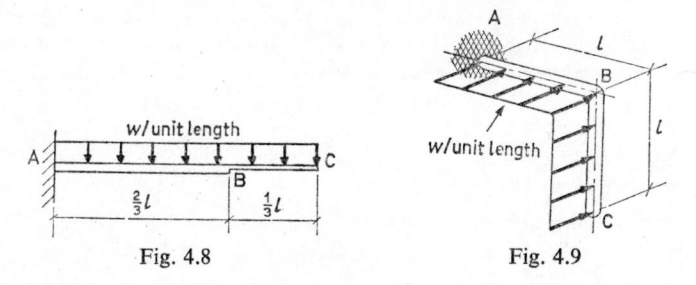

Fig. 4.8

Fig. 4.9

Problem 4.3 A cranked beam of circular section is shown in Fig. 4.9 loaded by a uniformly distributed load w per unit length. Determine the resultant deflection of C, twist at B in terms of w, l, E and I. (Note: $G = 0.4E$, $J = 2I$.)

Problem 4.4 Derive an expression for the horizontal deflection at C of the half-ring shown in Fig. 4.10. What magnitude of horizontal load would be required at C to push this point back to its original position?

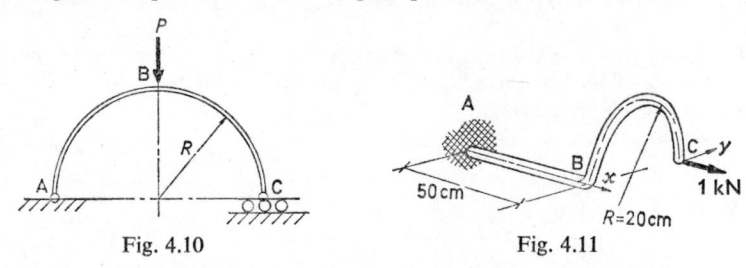

Fig. 4.10

Fig. 4.11

Problem 4.5 A 1 cm diameter rod is bent to the shape shown in Fig. 4.11 and fixed to a rigid support at A. Evaluate the deflection of point C in the direction of the applied load at this point. $E = 60$ kN/mm², $G = 25$ kN/mm².

Problem 4.6 Determine the horizontal and vertical deflections of point D for the beam shown in Fig. 4.12. Determine also the slope at D. Deduce the vertical deflection of B.

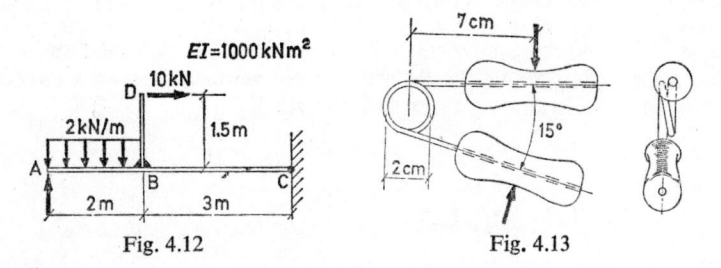

Fig. 4.12

Fig. 4.13

Problem 4.7 A spring stiffness of 30 N per cm of closure of the centre points of the handles is required for the wrist exerciser shown in Fig. 4.13. Determine a suitable diameter for the steel rod used to manufacture the exerciser. Consider bending effects only. $E = 200$ kN/mm².

Problem 4.8 The structure shown in Fig. 4.14 consists of a straight beam ABC of second moment of area I and a curved beam of second moment of area $\tfrac{1}{2}I$. Derive an expression for the deflection of C.

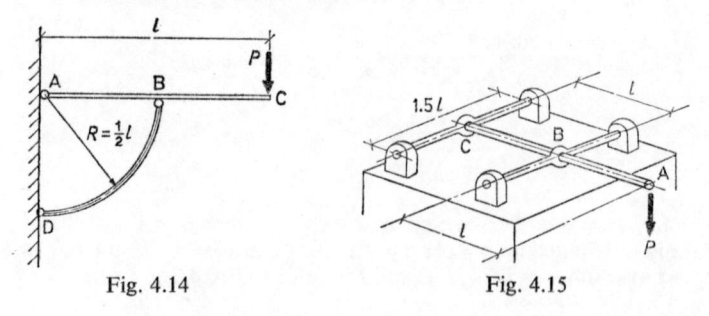

Fig. 4.14

Fig. 4.15

Problem 4.9 Assuming simple support conditions at all support points, determine the vertical deflections at A, B and C for the structure shown in Fig. 4.15. I for all members is the same.

Problem 4.10 Evaluate the central deflection of the cross-beam in the structure shown in Fig. 4.16. For the beams $E = 6 \text{ kN/mm}^2$, $I = 3 \times 10^6 \text{ mm}^4$. For the wire $E = 200 \text{ kN/mm}^2$, $A = 1 \text{ cm}^2$.

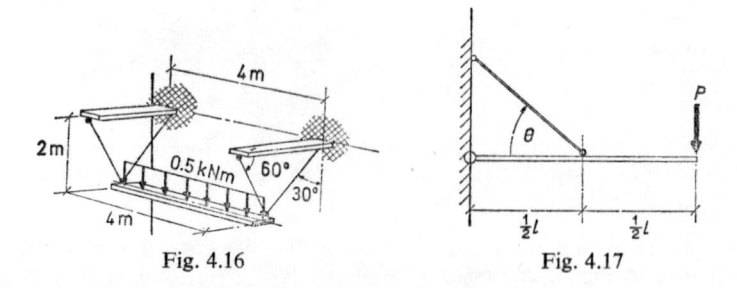

Fig. 4.16

Fig. 4.17

Problem 4.11 The structure shown in Fig. 4.17 consists of a beam of length l pinned at one end to a wall and joined to the wall also by a bar stretching from the beam centre point at an angle θ. Show that the load point deflection is a minimum if $\theta = 54.7°$. State this deflection in terms of P, E and I for the beam and A for the bar. Neglect direct load on the beam.

5

ANALYSIS OF STATICALLY INDETERMINATE STRUCTURES

5.1 Statically indeterminate beams

In this chapter we examine the application of energy principles to the analysis of statically indeterminate structures. We shall begin by studying beams in this section and proceed from there to analyse frame structures in the next section.

A very simple statically indeterminate beam is shown in Fig. 5.1. Here we have a propped cantilever, i.e. a cantilever with an additional support. The total number of support reactions which have to be evaluated is four, but we have only three equations of static equilibrium,

$$\Sigma P_x = \Sigma P_y = \Sigma M = 0,$$

and therefore a unique evaluation of the support reactions using equilibrium conditions alone is impossible.

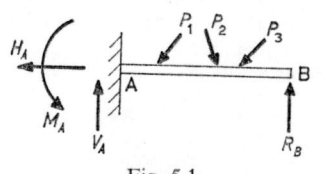

Fig. 5.1

Since the number of unknown forces is one more than the number of available equilibrium equations, the beam is considered to be statically indeterminate to the first degree. In general, the *degree of indeterminancy* is found by subtracting the number of relevant equilibrium equations from the total number of unknowns.

Now if any single reaction apart from H_A was removed from the beam the structure would then be statically determinate. Thus one of the reactions is said to be *redundant*, i.e. superfluous to the requirements for equilibrium. The number of redundant forces in a statically indeterminate structure is equal to the degree of indeterminancy.

In the analysis of statically indeterminate structures we obtain the additional equations required by examining the displacements of the

VIRTUAL WORK AND ENERGY CONCEPTS

structure. In general, if a redundant force acts on a structure it puts some constraint on the displacements. For example, the prop of Fig. 5.1 makes the tip deflection zero. Now, if we remove the prop the tip displaces by an amount Δ. The force in the prop must be sufficient to displace the tip back to its original position when the prop is replaced. Using this fact, we can evaluate the redundant force. The following examples illustrate the application of this method.

Example 5.1

Determine the reactions of the propped cantilever shown in Fig. 5.2.

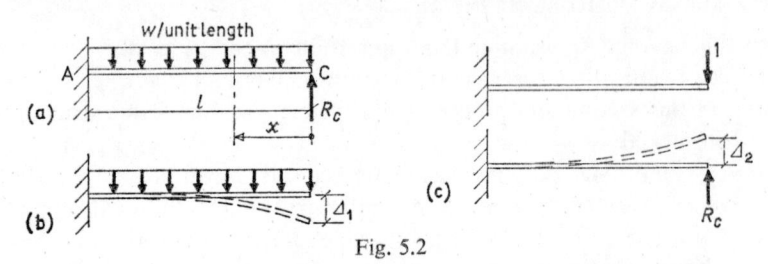

Fig. 5.2

Solution Here no applied horizontal forces act on the body so that equilibrium of horizontal forces requires $H_A = 0$ immediately. Two equilibrium equations remain to determine the three unknown reactions, so that the beam is singly indeterminate.

If we now remove one reaction, say the prop, we have a statically determinate structure, which we shall call the *primary* structure. The applied loading causes the tip of the cantilever to displace an amount Δ_1, this displacement being evaluated by applying a unit load at the tip.

That is
$$\Delta_1 = \int_0^l \frac{Mm}{EI} \, dx,$$

where
$$M = \tfrac{1}{2} w x^2, \quad m = x.$$

Therefore
$$\Delta_1 = \int_0^l \tfrac{1}{2} \frac{wx^3}{EI} \, dx = \tfrac{1}{8} \frac{wl^4}{EI}.$$

Now, if we replace the prop reaction R_C, the tip of the cantilever will displace by an amount Δ_2 where

$$\Delta_2 = \int_0^l \frac{M_R m}{EI} \, dx.$$

M_R is the moment caused by the prop reaction, and is proportional to m since both act at the same place, therefore $M_R = -R_C m$, the negative sign denoting that R_C acts in the opposite direction to the unit load.

Therefore
$$\Delta_2 = -R_C \int_0^l \frac{m^2}{EI} \, dx = -R_C \int_0^l \frac{x}{EI} \, dx = -R_C \frac{l^3}{3},$$

STATICALLY INDETERMINATE STRUCTURES

where Δ_2 is measured positive down as is Δ_1. The total displacement of the tip is obtained by superposing the displacements of the primary structure due to the applied loading and due to the redundant reaction. That is

$$\Delta_{TIP} = \Delta_1 + \Delta_2 = \frac{wl^4}{8EI} - \frac{R_C l^3}{3}.$$

Now since the prop does not displace we have $\Delta_{TIP} = 0$, from which

$$R_C = \tfrac{3}{8}wl.$$

Knowing R_C, it is a simple matter to apply the equilibrium equations to obtain the other two reactions.

$\Sigma P_y = 0$ gives $R_A - wl + \tfrac{3}{8}wl = 0$, whence $R_A = \tfrac{5}{8}wl$

$\Sigma M_A = 0$ gives $M_A - \tfrac{1}{2}wl^2 + \tfrac{3}{8}wl^2 = 0$, whence $M_A = \tfrac{1}{8}wl^2$.

This method of approach is applicable to all statically indeterminate beam problems. Any of the support reactions could have been chosen as the redundant reaction with the same final results.

Example 5.2

The beam shown in Fig. 5.3 is simply supported at both ends and an intermediate support is supplied through a spring, stiffness K, at the centre. Determine the force in the spring when the beam is loaded by a uniformly distributed load of w/unit length.

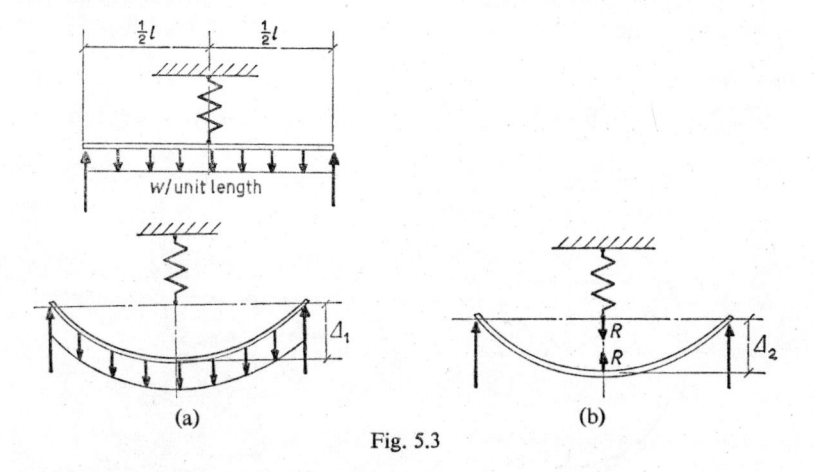

Fig. 5.3

Solution In this case the structure is composed of two members, the beam and the spring. We shall choose the spring force as the redundant reaction on the beam. To remove this force we disconnect the spring from the support and obtain the primary structure as shown in which the applied load is carried

56 VIRTUAL WORK AND ENERGY CONCEPTS

completely by the beam and the spring is unloaded. Under the applied loading the beam centre deflection is now obtained by applying a unit load at the centre.

$$\Delta_1 = 2 \int_0^{\frac{1}{2}l} \frac{Mm}{EI} dx = \frac{2}{EI} \int_0^{\frac{1}{2}l} \left(\frac{wlx}{2} - \frac{wx^2}{2} \right) \frac{x}{2} dx = \frac{5}{384} \frac{wl^4}{EI}.$$

The evaluation of M and m in this case is as shown in Example 4.1. We must now apply the spring force R to the spring as shown in order to re-connect spring to the support. The displacement of the spring is R/K and the displacement of the beam is

$$\Delta_B = \frac{2l}{EI} \int_0^{\frac{1}{2}l} \left(\frac{x}{2} \right)^2 dx = \frac{1}{48} \frac{Rl^3}{EI}.$$

The total closure of the gap is therefore

$$\Delta_2 = \left(\frac{R}{K} + \frac{1}{48} \frac{Rl^3}{EI} \right).$$

Since spring and support must re-connect, $\Delta_1 = \Delta_2$. Therefore

Hence
$$\frac{5}{384} \frac{wl^4}{EI} = R \left(\frac{1}{K} + \frac{1}{48} \frac{l^3}{EI} \right)$$

$$R = \frac{\dfrac{5}{384} \dfrac{wl^4}{EI}}{\left(\dfrac{1}{K} + \dfrac{1}{48} \dfrac{l^3}{EI} \right)} = \frac{5wl}{8 + \dfrac{384EI}{Kl^3}}.$$

Note: as $K \to \infty$, $R \to \tfrac{5}{8}wl$, this is the reaction at the middle support if the beam is simply supported at centre and both ends. Also, if $K = 0$, $R = 0$, i.e. we have a beam simply supported at both ends with no intermediate support.

Example 5.3

Derive an expression for the bending moment at any point of a thin ring under the loading shown in Fig. 5.4.

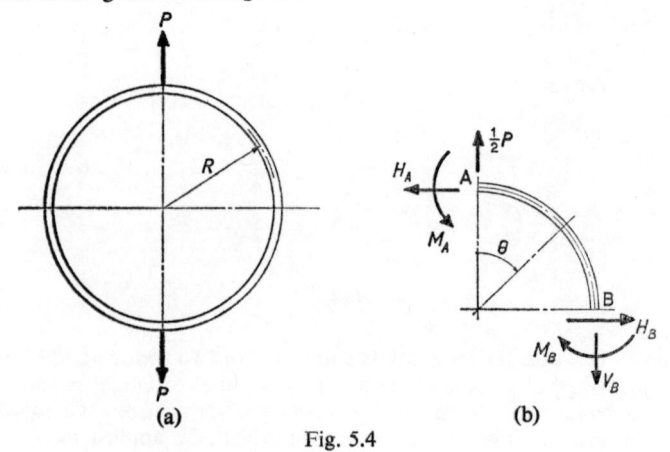

Fig. 5.4

STATICALLY INDETERMINATE STRUCTURES

Solution To analyse the ring we consider one quarter, noting that from symmetry the loading on each quarter is the same. Since shear force is zero or passes through zero at a point of symmetry $H_B = 0$. $\Sigma Px = 0$ therefore gives $H_A = 0$. $\Sigma P_y = 0$ gives $V_B = \frac{1}{2}P$.

We now have two unknowns M_A and M_B and only one equation of equilibrium left, so the ring is singly indeterminate.

We shall consider M_A redundant and remove this force. We must now obtain the rotation at A (i.e. the displacement corresponding to M_A) caused by the loading on the primary structure. We do this by applying a unit moment clockwise at A. We have $M = \frac{1}{2}Pr \sin \theta$, $m = 1$;

$$\theta_1 = \int \frac{Mm}{EI} \, dx = \int_0^{2\pi} \frac{Mm}{EI} r \, d\theta = \frac{4}{EI} \int_0^{\frac{1}{2}\pi} \tfrac{1}{2}Pr \sin \theta \times 1r \, d\theta.$$

Therefore

$$\theta_1 = \frac{2Pr^2}{EI}.$$

Now, applying the redundant force M_A, we have

$$\theta_2 = -M_A \int_0^{2\pi} \frac{m^2}{EI} r \, d\theta.$$

That is

$$\theta_2 = -\frac{4M_A}{EI} \int_0^{\frac{1}{2}\pi} 1^2 r \, d\theta = -M_A \times \frac{2\pi r}{EI}.$$

Because of symmetry the rotation of the ring at A must be zero. Therefore $\theta_1 + \theta_2 = 0$.

Hence

$$M_A = \frac{Pr}{\pi}.$$

The required expression for bending moment at any point is

$$M = \tfrac{1}{2}Pr \sin \theta - M_A = \tfrac{1}{2}Pr \left(\sin \theta - \frac{2}{\pi} \right).$$

Example 5.4

Determine the support reactions for the built-in beam shown in Fig. 5.5 (a) and draw the bending moment diagram.

Solution In this case we have four unknown reactions and only two useful equilibrium equations. The structure is indeterminate to the second degree. We therefore have two redundant reactions which we must remove to obtain the primary structure. We shall choose R_C and M_C as the redundant reactions and remove them from the structure leaving a cantilever once more as shown in Fig. 5.5 (b). For the cantilever $M = x^2$ for $x < 2$ and $M = x^2 + 6(x-2)$ for $2 < x < 3$.

Since we have removed two forces we must examine both corresponding displacements θ_C and Δ_C. We therefore apply a unit load downwards and a unit moment clockwise at C. The unit load gives $m_1 = x$, and the unit rotation gives

VIRTUAL WORK AND ENERGY CONCEPTS

$m_2 = 1$. The moment caused by the reaction force R_C is therefore $-R_C m$, and that caused by M_C is $+M_C m_2$.

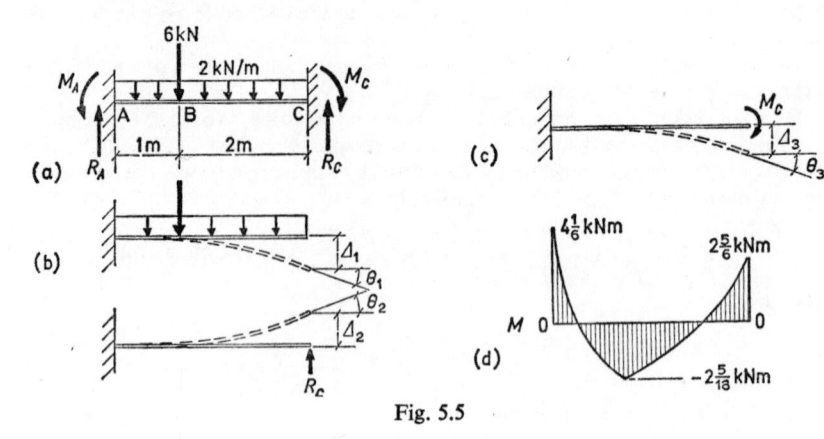

Fig. 5.5

Due to the applied loading on the primary structure the displacements are

$$\Delta_1 = \int \frac{Mm_1}{EI} \, dx = \frac{1}{EI} \int_0^2 x^2 x \, dx + \frac{1}{EI} \int_2^3 [x^2 + 6(x-2)] x \, dx$$

$$= \frac{1}{EI} \{4 + [\tfrac{1}{4}x^4 + 2x^3 - 6x^2]_2^3\} = \frac{28.25}{EI}.$$

Similarly

$$\theta_1 = \frac{1}{EI} \int Mm_2 \, dx = \frac{1}{EI} \{\tfrac{8}{3} + [\tfrac{1}{3}x^3 + 3x^2 - 12x]\} = \frac{1.2}{EI}.$$

Due to the reaction R_C we have also a displacement and a rotation at C, given by

$$\Delta_2 = -R_C \int \frac{m_1^2}{EI} \, dx = -\int_0^3 \frac{R_C}{EI} x^2 \, dx = -\frac{9R_C}{EI}$$

$$\theta_2 = -R_C \int \frac{m_1 m_2}{EI} \, dx = -\frac{R_C}{EI} \int_0^3 x \, dx = \frac{4.5R_C}{EI}.$$

Similarly we have, due to M_C

$$\Delta_3 = \frac{M_C}{EI} \int m_1 m_2 \, dx = \frac{4.5M_C}{EI}$$

$$\theta_3 = \frac{M_C}{EI} \int m^2 \, dx = \frac{3M_C}{EI}.$$

Since both the net displacement and rotation of point C are zero we have

$$\Delta_1 + \Delta_2 + \Delta_3 = 0$$

$$\theta_1 + \theta_2 + \theta_3 = 0.$$

STATICALLY INDETERMINATE STRUCTURES

Substituting for θ and Δ gives the simultaneous equations

$$9R_C - 4.5M_C = 28.25$$
$$4.5R_C - 3M_C = 12.$$

Solving, we have $R_C = 4\frac{5}{8}$ kN, $M_C = 2\frac{5}{6}$ kN/m.

To draw the bending moment diagram we simply apply R_C and M_C to the primary structure together with the applied loading and analyse as a statically determinate beam. The resulting bending moment diagram is shown in Fig. 5.5 (d).

5.2 Statically indeterminate frameworks

The same basic approach of removing redundant forces to obtain the statically determinate primary structure is used in the analysis of statically indeterminate frameworks. In such structures the evaluation of support reactions does not always enable us to obtain the forces in all members. Consider for example the frame shown in Fig. 5.6. Here we see that the

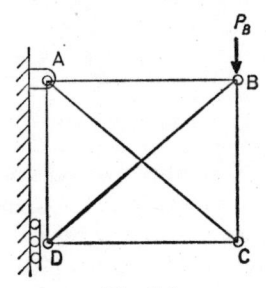

Fig. 5.6

external reactions are statically determinate, but we cannot obtain the forces in the bars. The frame is therefore *internally redundant*. If we remove any single bar the bar forces in the rest of the structure can be evaluated statically, therefore the frame is singly redundant.

The degree of indeterminancy in frame structures is not always easy to see immediately. Because of this it is wise to use a set rule for obtaining the degree of redundancy. We obtain this rule by considering the number of unknown forces and the number of available equilibrium equations. The total number of forces is found by adding the number of members and the number of reactions, i.e. $n+r$. The total number of equilibrium equations is obtained by applying the equilibrium conditions to the horizontal and vertical components of force at each joint, giving $2j$ equations. The degree of redundancy of the frame is therefore equal to $n+r-2j$ where n is the number of members, r is the number of unknown reactions and j is the number of joints.

If we remove one member from the structure of Fig. 5.6, say member AB and obtain the primary structure we can obtain the forces in each bar due to the applied loading. The increase in distance between joint A and joint B is found by applying unit loads at A and B as shown in Fig. 5.7 and evaluating Δ_1 from

$$\Delta_1 = -\sum \frac{Ppl}{AE}.$$

The negative sign is used since Δ_1 denotes an increase in distance whereas the unit loads produce a decrease in distance.

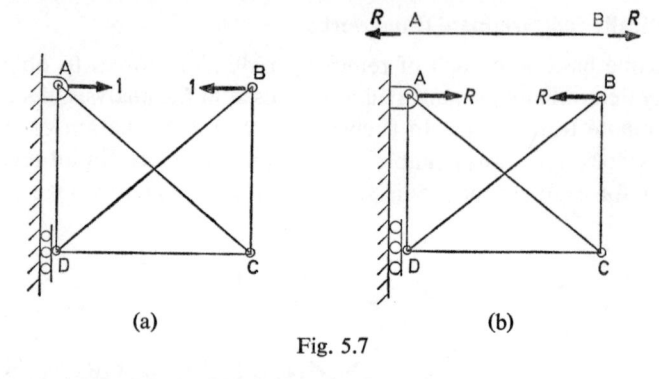

Fig. 5.7

If we now try to re-connect member AB to the primary structure we have to apply equal forces to both ends of bar AB and to joints A and B as shown. Denoting the applied force by R the extension of the bar AB is obtained by applying a unit load to the bar giving

$$\Delta_{bar} = \frac{R \times 1 \times l_{AB}}{AE}.$$

The closure of joints A and B due to the redundant force R is

$$\Delta_{frame} = R \sum \frac{p^2 l}{AE}.$$

The closure of the joints of the primary structure plus the extension of bar AB must equal the increase in distance between joints A and B of the primary structure if the bar is to be re-connected, we therefore have

$$\frac{Rl_{AB}}{AE} + R \sum \frac{p^2 l}{AE} = -\sum \frac{Ppl}{AE}.$$

We can combine the two expressions on the left-hand side of equation by including the redundant member in the summation, giving

$$R \sum \frac{p^2 l}{AE} = -\sum \frac{Ppl}{AE},$$

STATICALLY INDETERMINATE STRUCTURES

where the summation includes all members. From this we obtain

$$R = -\frac{\sum \frac{Ppl}{AE}}{\sum \frac{p^2 l}{AE}}.$$

The application of this equation is shown in the following example.

Example 5.5

Obtain the bar forces in the structure shown in Fig. 5.8. All members have cross-sectional area of 200 mm² and Young's Modulus $E = 200$ kN/mm².

Fig. 5.8

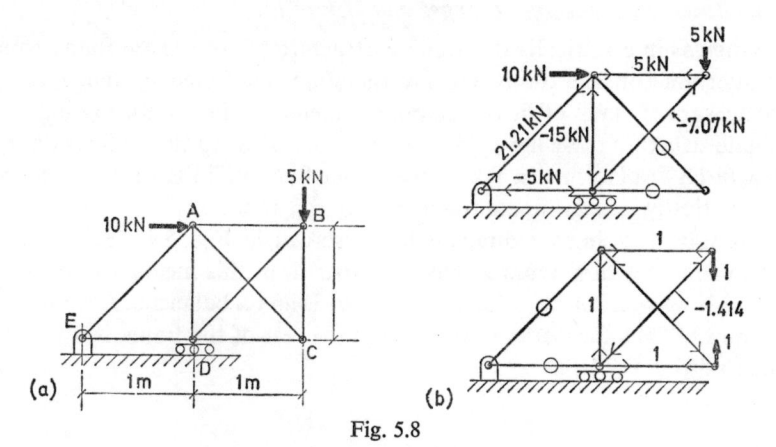

Solution We choose member BC as the redundant member and remove it to obtain the primary structure The bar forces in the primary structure are shown opposite as are the bar forces due to unit loads at joints B and C. In evaluating the required summations we shall use a tabular method as shown. Since E and A for the members is constant, we can omit these from the table and evaluate only ΣPpl and $\Sigma p^2 l$.

Member	P	p	Ppl	p²l	P total (= P + Rp)
AB	5	1	5	1	+4.57
AC	0	−1.414	0	2.828	+0 16
AD	−15	1	−15	1	−15.43
AE	21.21	0	0	0	+21.21
BD	−7.07	−1.414	14.14	2.828	−6.46
DC	0	1	0	1	−0.43
DB	−5	0	0	0	−5
BC	0	1	0	1	−0.43
			Σ = 4.14	Σ = 9.656	

$$R = -\frac{\sum \dfrac{Ppl}{AE}}{\sum \dfrac{p^2 l}{AE}} = -\frac{4.14}{9.656} = -0.439 \text{ kN}.$$

The total force in each member is obtained by superposing the forces obtained from the primary structure and those due to the redundant force R.

That is $$P_{total} = P + R \times p$$

These are shown in the table.

The effects of temperature changes and lack of fit

Whereas in a statically determinate structure the forces are found from equilibrium considerations and are therefore unaffected by temperature changes, or by lack of fit of component members, in a statically indeterminate structure these have effects on the forces due to their effects on the structure's displacements. We shall examine first of all the stresses induced in a statically indeterminate frame due to lack of fit.

Consider the singly redundant frame shown in Fig. 5.9. Member AB is too short by an amount λ. Now in order to fit this member into joints A and B we must apply a force R_T at joint B and a balancing force to AB as shown. Thus the bar is stretched and the rest of the frame is distorted due to fitting the bar.

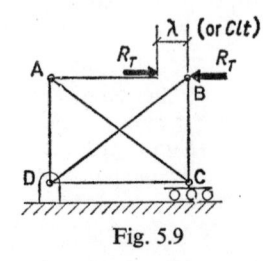

Fig. 5.9

The forces induced in the members can be evaluated if the value of R_T required to close the gap is known. We therefore apply unit loads at the end of member AB and at joint B, the deflection between these points being

$$\Delta \times 1 = \sum \frac{Ppl}{AE}.$$

Now we know that the unit loads are in the same direction as the applied forces R_T, so that the forces p are in proportion to P,

i.e. $$P = R_T p.$$

STATICALLY INDETERMINATE STRUCTURES

Also Δ the displacement must be equal to λ if the gap is to be closed so we can write

$$\lambda = R_T \sum \frac{p^2 l}{AE}.$$

Rearranging gives

$$R_T = \frac{\lambda}{\sum \dfrac{p^2 l}{AE}}.$$

Now if the frame is acted upon by external loads the bar forces are found by adding the forces due to lack of fit to those due to the loading.

Example 5.6

If the member BC of the frame of Example 5.5 is made 2 mm short, estimate the forces in all members due to assembly and the total forces in the members under load.

Solution From the Example 5.5, we know the member forces due to both the applied loading and due to unit loads at B and C. We can therefore evaluate the forces due to lack of fit immediately and add these to the forces due to the loading to obtain the total member forces.

From Example 5.5, $\quad \sum p^2 l = 9.656$ m

$$= 9.656 \times 10^3 \text{ mm}.$$

Therefore

$$\sum \frac{p^2 l}{AE} = 9.656 \times 10^3 / 200 \times 200$$

$$= 0.242 \text{ mm/kN}.$$

Hence

$$R_T = \frac{\lambda}{\sum \dfrac{p^2 l}{AE}} = \frac{2}{242} = 8.25 \text{ kN}.$$

The total load in each member is given by $P_T = P + R_T p$. These loads are tabulated below:

Member	AB	AC	AD	AE	BD	DC	DE	BC
P	−4.51	0.61	−15.43	21.21	−6.46	−0.43	−5	−0.43
$P+pR_T$	3.68	−11.07	−7.18	21.21	−18.14	7.82	−5	8.25

With regard to temperature effects we will consider once more the frame of Fig. 5.9 assuming for the purpose of clarity that there is no lack of fit in this instance. Now if part of the structure is subjected to a temperature change this part will tend to deform and these deformations will be resisted by the other members thus setting up forces in the members. For the frame shown, if the temperature of the bar AB is reduced by $t°$ it

64 VIRTUAL WORK AND ENERGY CONCEPTS

would shorten by an amount clt (as shown in Fig. 5.9) if it were released from joint B; c is the coefficient of linear expansion of the material.

Since we must now close the gap which has arisen, we are faced with a a problem analogous to the lack of fit problem, and this is tackled in exactly the same way, with the quantity clt being substituted for λ. In this way we arrive at the expression for the force in the bar AB,

$$R_T = \frac{clt}{\sum \dfrac{p^2 l}{AE}}.$$

Problems involving temperature effects are therefore tackled in exactly the same way as lack of fit problems. Additional bar forces due to temperature effects can be added to those due to external loading and lack of fit.

5.3 Deflections of statically indeterminate structures

Suppose we have performed the analysis of the propped cantilever of Example 5.1 and now wish to determine the centre displacement. If we apply a unit load at the centre of the beam, it would appear that we cannot immediately obtain the variation of bending moment along the beam due to the unit load, since the beam is statically indeterminate. To clarify this point let us examine the virtual complementary work due to a virtual load on such a beam as shown in Fig. 5.10, displaced under some external

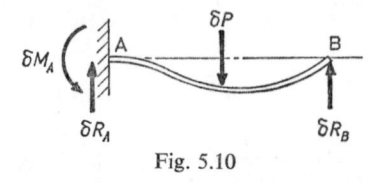

Fig. 5.10

force action. The complementary work done by the virtual force is $\Delta \delta P$, where Δ is the actual displacement of the structure and δP is the virtual load. The complementary work done by the virtual reactions δM_A, δR_A and δR_B is $\theta_A \delta M_A$, $\Delta_A \delta R_A$ and $\Delta_B \delta R_B$ respectively. Since θ_A, Δ_A and Δ_B are zero, it follows that no complementary work is done by these reactions. Therefore, if we wish to remove any reaction, we can do so without altering the virtual complementary work situation owing to a virtual force. We can use this fact to ease the labour involved in computing deflections of statically indeterminate structures as follows. In determining the virtual loading system (i.e. m, t, etc.) we remove constraints to obtain a statically determinate structure and apply the unit load to this structure thus obtaining m, t, etc., very simply. For the real loading we must of course use the actual moments, etc., present on the indeterminate structure.

STATICALLY INDETERMINATE STRUCTURES

Two examples are shown here to illustrate the method of obtaining displacements.

Example 5.7

Determine the centre displacement of the propped cantilever of Example 5.1.

Solution From Example 5.1 the moments on the beam at any point are given by
$$M = R_C x - \tfrac{1}{2}wx^2 \quad \text{and} \quad R_C = \tfrac{3}{8}wl.$$
Therefore
$$M = w(\tfrac{3}{8}xl - \tfrac{1}{2}x^2).$$

Now to obtain the displacement Δ_C we apply a unit load at the centre. Since we can remove reactions to make the structure statically determinate and simplify the analysis we shall take away the prop. The bending moment m at any section is in Fig. 5.11.
$$m = 0 \text{ for } x < \tfrac{1}{2}l \text{ and } m = \tfrac{1}{2}l - x \text{ for } x > \tfrac{1}{2}l.$$

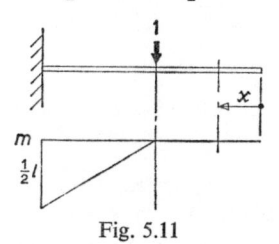

Fig. 5.11

Using the unit load equation we have
$$\Delta_C = \int_0^l \frac{Mm}{EI} dx = \int_0^{\frac{1}{2}l} 0\, dx + \frac{1}{EI}\int_{\frac{1}{2}l}^l w(\tfrac{3}{8}xl - \tfrac{1}{2}x^2)(\tfrac{1}{2}l - x)\, dx.$$
Integrating gives
$$\Delta_C = \frac{w}{EI}\left[\tfrac{1}{8}x^4 - \tfrac{5}{24}x^3 l + \tfrac{3}{32}x^2 l^2\right]_{\frac{1}{2}l}^l = \tfrac{1}{192}\frac{wl^4}{EI}.$$

Note that the same result would be obtained if we removed one of the other reactions instead of the prop.

Example 5.8

Determine the vertical displacement of joint B for the structure of Example 5.5.

Solution From Example 5.5 we have the forces in the members. These are tabulated in that example. To find the required displacement we apply a unit vertical load at B. We can reduce the structure to a statically determinate one by removing any member we choose. We shall select member BC to remove. The forces in the members due to the unit load at B are shown in Fig. 5.12. The vertical displacement of point B is

$$\Delta_B = \sum \frac{Ppl}{AE} = \frac{10^3}{AE}[P_{AB} \times 1 + P_{BD} \times 1.414 \times (-1.414)$$
$$+ P_{AE} \times 1.414 \times 1.414 + P_{AD} \times (-1) + P_{ED} \times (-1)].$$

66 VIRTUAL WORK AND ENERGY CONCEPTS

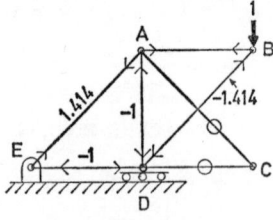

Fig. 5.12

Substituting the relevant values from Example 5.5, we have

$$\Delta_B = \frac{10^3}{200 \times 200} [4.57 - 6.46 \times (-1.414)$$
$$\times 1.414 \times 21.21 \times 1.414 \times 1.414 - 15.43 \times (-1) - 5 \times (-1)],$$
$$= 2 \text{ mm}.$$

Problems

Problem 5.1 Check the answers obtained for Examples 5.2 to 5.5 by considering different forces to be redundant than those used previously.

Problem 5.2 A beam is simply supported at both ends and at its centre point. A load is applied which varies linearly in intensity from zero at one end to w per unit length at the other end. Obtain the values of the support reactions.

Problem 5.3 A beam is built in at both ends. A uniformly distributed load of intensity w per unit length acts over one half of the beam, from the beam centre to one end. Evaluate the support reactions.

Problem 5.4 Evaluate the support reactions for the wall fixture shown in Fig. 5.13.

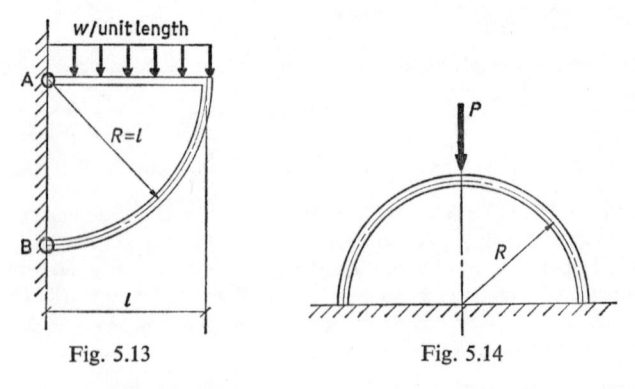

Fig. 5.13   Fig. 5.14

Problem 5.5 The half ring shown in Fig. 5.14 is rigidly fixed at both ends. Determine the maximum moment due to a vertical load P. Evaluate the displacement of the load point.

STATICALLY INDETERMINATE STRUCTURES

Problem 5.6 In the clamp shown in Fig. 5.15, the screw has one thread per millimetre, i.e. one turn of the handle moves the screw one millimetre through the anvil. Considering only the effects of bending of the anvil, evaluate the clamping force produced on a rigid body per turn of the handle. $E = 200$ kN/mm².

Fig. 5.15

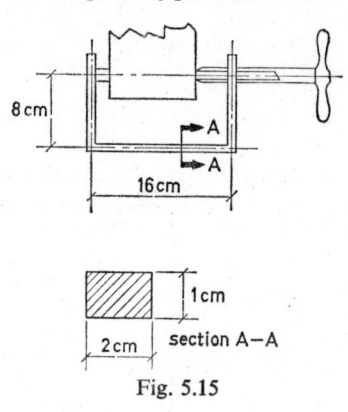

Problem 5.7 Determine the member forces in the frame structures shown in Fig. 5.16. Evaluate the horizontal displacement of joint H in Fig. 5.16 (a) and the vertical displacement of joint E in Fig. 5.16 (b).

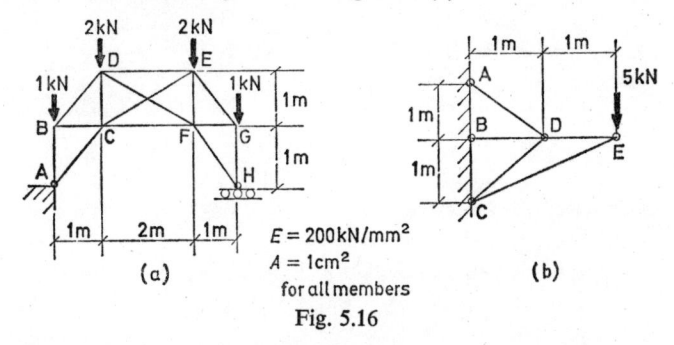

$E = 200$ kN/mm²
$A = 1$ cm²
for all members

Fig. 5.16

Problem 5.8 In the assembly of the structure shown in Fig. 5.17, it was found that after fixing the support at A the point C is 2 mm vertically away from its mating

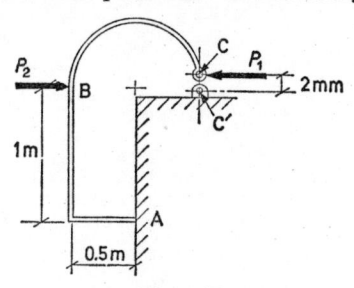

Fig. 5.17

point C'. Both C and C' lie on the same vertical line. In order to bring points C and C' into coincidence, and thus complete the assembly, it is only possible to apply forces horizontally at C and B as shown. Determine the required magnitudes of P_1 and P_2. $E = 200$ kN/mm², $I = 1.2 \times 10^6$ mm⁴.

Problem 5.9 When the structure of Problem 5.8 has been pinned at C the loads P_1 and P_2 are removed. What reaction forces are now present?

Problem 5.10 Evaluate the centre displacement of the beam of Problem 5.3.

Problem 5.11 The structure shown in Fig. 5.18 consists of a beam pinned at its end and supported also by two bars. If the bars undergo a temperature rise of 50°C, what forces are induced in them.

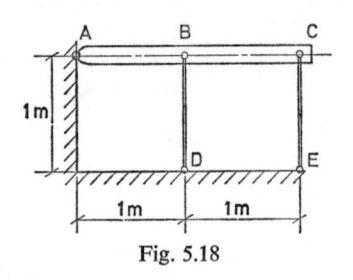

Fig. 5.18

For the beam $E = 207$ kN/mm², $I = 4 \times 10^6$ mm⁴
For the bars $E = 207$ kN/mm², $A = 100$ mm².
Coefficient of linear expansion, $C = 12 \times 10^{-6}$ m per m per °C.

ANSWERS TO PROBLEMS

Chapter 1
1.1 $h_1 = 0.711l$ 1.2 $\theta = 18.3°$

1.3 $M = Wl \sin \theta$ 1.4 $P = \dfrac{W}{\dfrac{\frac{3}{2}-\cos\theta}{\sqrt{1-(\frac{3}{2}-\cos\theta)^2}} - \cot\theta}$

1.5 $P = \frac{1}{2}W\left(\dfrac{2a}{b} - 1\right)\tan\frac{1}{2}\theta$ 1.6 $w = \dfrac{4Pa\sqrt{a^2 - \left(\dfrac{x}{2}\right)^2}}{x(a+b)}$

1.7 $P = 0.465$ kN.

Chapter 2
2.1 $P = KR(\frac{2}{3}\sin\theta - \sin\frac{1}{2}\theta)$ 2.2 $\theta = 37.5°$

2.3 $P = 444$ N   2.4 $h = 2\left(\dfrac{2w}{K} - l\right)$ for $w > \frac{1}{2}Kl$

2.5 $\Delta_A = 5\frac{5}{13}$ cm to left, $P = 30.77$ N   2.6 $P_1 = 1.4$ kN, $P_2 = 2.6$ kN

2.7 $\theta = \dfrac{K_2}{4K_1 + K_2} \cdot \dfrac{\pi}{2}$, $P = \dfrac{K_2\left(\dfrac{\pi}{2} - \theta\right)}{l\cos\theta}$ 2.8 $\Delta_A = \dfrac{28P}{K}$.

Chapter 3
3.1 $\Delta_B = 2.87$ mm
3.2 $\Delta_V = 2.85$ mm, $\Delta_H = 54.27$ mm
3.3 $\Delta_A = 8$ mm, $\Delta_B = 4.5$ mm
3.4 Δ_B (minimum) $= 2.5$ mm. Δ_A does not change
3.5 $\Delta_H = 20.5$ mm, $\Delta_V = 44.1$ mm
    Resultant deflection is 48.6 mm in direction 65° from horizontal.

Chapter 4

4.1 $y_C = \dfrac{P}{EI}\left[\tfrac{1}{16}lX^2 - \tfrac{1}{12}X^3\right]$ $X < \tfrac{1}{2}l$ 4.2 $y = \tfrac{41}{648}\dfrac{wl^4}{EI}$

4.3 $\Delta_B = \tfrac{29}{24}\dfrac{wl^4}{EI}$, $\theta_C = \tfrac{5}{8}\dfrac{wl^3}{EI}$ 4.4 $\Delta_C = \tfrac{1}{2}\dfrac{PR^3}{EI}$, load required at $C = \dfrac{P}{\pi}$

4.5 4.65 cm

4.6 Horizontal deflection of $D = 22.5$ mm, vertical deflection of $D = 11.6$ mm, rotation of $D = 1.02°$, vertical deflection of $B = 11.6$ mm

4.7 3.8 mm

4.8 $\Delta_A = 1.018\dfrac{Pl^3}{EI}$, $\Delta_B = \tfrac{1}{24}\dfrac{Pl^3}{EI}$, $\Delta_C = -\tfrac{1}{48}\dfrac{Pl^3}{EI}$

4.9 $\Delta_C = 0.225\dfrac{Pl^3}{EI}$

4.10 150 mm

4.11 $\Delta = \dfrac{P}{E}\left[\tfrac{1}{12}\dfrac{l^3}{I} + \dfrac{5.25l}{A}\right]$.

Chapter 5

5.2 $\tfrac{1}{96}wl$, $\tfrac{5}{16}wl$, $\tfrac{17}{96}wl$

5.3 $\tfrac{13}{32}pl$, $\tfrac{3}{32}pl$. Support moments are $\tfrac{11}{192}pl^2$, $\tfrac{5}{192}pl^2$

5.4 $V_B = 0.405wl$, $V_A = 0.595wl$, $_A = -H_B = \tfrac{1}{2}wl$

5.5 $M_{max} = 0.151PR$, $\Delta_P = 0.0133\dfrac{PR^3}{EI}$

5.6 244 N

5.7 (a) Sample forces:
$P_{CF} = -P_{DE} = 2$ kN. $P_{CE} = P_{DF} = P_{EF} = P_{DC} = P_{AC} = P_{FH} = 0$. $\Delta_H = 1.665$ mm

(b) $P_{AD} = P_{CD} = 2.93$ kN. $P_{BD} = 5.85$ kN. $\Delta_E = 4.435$ mm

5.8 $P_1 = 1.05$ kN, $P_2 = 1.3$ kN

5.9 $R_H = 0.195$ kN, $R_V = 0.321$ kN

5.10 $\tfrac{1}{768}\dfrac{\phi l^4}{EI}$

5.11 $F_{CE} = 0.572$ kN, $F_{BD} = 1.144$ kN.

INDEX

Axial Forces 35, 39

Bars 35, 39
Beam, bending moment diagram 57, 58
  built in 57, 58
  cantilever 28, 44, 50, 51
  cranked 46, 51
  curved 46, 47, 51
  deflections 43
  differential equation of 7
  propped cantilever 53, 54, 64, 65
  simply supported 43, 45, 50
  statically determinate 43
  statically indeterminate 53
Bending, simple 7, 35
  equations 36
  moment diagram 57, 58
  virtual complementary energy of 36
Bernoulli, J. 7
Body, rigid 10
  elastic 16
Buckling load 22

Cantilever beam 28, 44, 50, 51
Castigliano, A. 7, 29
Circular ring 56
Clapeyron 7
Combined effects 38
Compatability 22
Complementary, strain energy 17
  virtual strain energy 17
  virtual work 26
  work 16
Cranked beams 44
Curved beams 46, 47
Cylinders 13
Cylindrical ring 56

Da Vinci, Leonardo 7
Deflections, of beams 43
  of frames 38
  due to bending 43
  due to shear 43
  due to torsion 46, 47
  of statically determinate structures 34
  of statically indeterminate structures 64, 65, 66
  of structures with several beam and bar members 47
Degrees of freedom 22
Degree of Indeterminancy 53, 59

Elastic Bodies 15, 16
Elasticity, Modulus of 19
Engesser, F. 7
Equilibrium, Principle of 7

Fixed Supports 43
Force-displacement diagram 16
Forces, equilibrium of 7
  work done by 8
Frame structures, deflections of 38
  statically determinate 38
  statically indeterminate 59

Galileo, G. 7

Ideal systems 11
Internal redundancy 59

Joints, equilibrium of 39, 59

Kinetic Energy 15

Lack of Fit 52, 63
Lagrange, J. L. 7

Mechanisms 10, 11, 14
Modulus, of elasticity 19
  shear 37
Mohr, O. 7
Moment, work done by 8

Neutral axis 35
Neutral position for springs 20, 21

Pin-jointed frames 38, 39
Primary structure 54
Principle, of equilibrium 7
 of virtual complementary work 26
 of virtual displacements 8, 9
 of virtual work 7, 8, 9, 20
Propped cantilever 53, 54, 64, 65

Redundant reactions 53, 54
Redundancy, degree of 53, 59
Rigid member structures 10

Second moment of area 36
Shear, deflections 2
 equations 36
 force 36, 37
 Modulus 37
 strain 19
 stress 19, 36, 37,
 virtual complementary energy of 36
Simple bending 7
Simply supported beams 43, 45, 50
Spring, systems 16
 stiffness 18
Statically determinate beams 43
 frames 38
Statically indeterminate beams 53
 frames 59
Strain 7
Strain, direct 18
 shear 19
Strain Energy 15, 17
 of springs 18
 of elastic bodies 19
Stress, direct 18
 shear 19
Structures with several members 47
Supports 39, 43, 53
Support reactions 53, 56, 57

Systems, ideal 11
 with one degree of freedom 22
 with many degrees of freedom 22

Tabular method of frame analysis 41, 61
Temperature effects 62, 64
Theorem of Castigliano 29, 30
Torsion, constant 37
 equations 37
 springs 21, 33
 virtual complementary energy of 38
Tyne, J. R. 7

Unit displacement method 29
Unit load, equation 28, 34, 43
 method 27
Unit Moment 44, 45
 Torque 44

Virtual complementary work, Principle of 26
Virtual, displacements 8, 9, 11, 16, 23
 loads 17
 rotations 10, 11, 25
 work 8, 16, 19, 20
Virtual complementary strain energy 35
 due to bending 36
 due to direct force 35
 due to shear 37
 due to torsion 38
Virtual work, Principle of 7

Walker, A. C. 38
Work done, by a force 8
 by a moment 8

Young's Modulus of Elasticity 19